The Bridge to Dark Matter; A New Sister Universe; Dark Energy; Inflatons; Quantum Big Bang; Superluminal Physics; An Extended Standard Model Based on Geometry

STEPHEN BLAHA, PH. D.

BLAHA RESEARCH

Blaha Research, P. O. Box 368, Auburn, NH 03032 USA

ISBN: 978-0-9845530-9-9

Cover Credits

A 19th Century photograph of a walking bridge in the garden of the Dowager Empress of China. © Copyright 2013 Stephen Blaha All Rights Reserved.
Cover design by Stephen Blaha © Copyright 2013. All Rights reserved.

Rev. 00/00/01 June 11, 2013

To My Wife Margaret

Some Other Books by Stephen Blaha

From Asynchronous Logic to The Standard Model to Superflight to the Stars (Blaha Research, Auburn, NH, 2011)

From Asynchronous Logic to The Standard Model to Superflight to the Stars volume 2: Superluminal CP and CPT, U(4) Complex General Relativity and The Standard Model, Complex Vierbein General Relativity, Kinetic Theory, Thermodynamics (Blaha Research, Auburn, NH, 2012)

The Algebra of Thought & Reality: The Mathematical Basis for Plato's Theory of Ideas, and Reality Extended to Include A Priori Observers and Space-Time; Second Edition (Pingree-Hill Publishing, Auburn, NH, 2009)

The Origin of the Standard Model: The Genesis of Four Quark and Lepton Species, Parity Violation, the ElectroWeak Sector, Color SU(3), Three Visible Generations of Fermions, and One Generation of Dark Matter with Dark Energy (Pingree-Hill Publishing, Auburn, NH, 2007)

Physics Beyond the Light Barrier: The Source of Parity Violation, Tachyons, and A Derivation of Standard Model Features (Pingree-Hill Publishing, Auburn, NH, 2007)

Quantum Theory of the Third Kind: A New Type of Divergence-free Quantum Field Theory Supporting a Unified Standard Model of Elementary Particles and Quantum Gravity based on a New Method in the Calculus of Variations (Pingree-Hill Publishing, Auburn, NH, 2005)

Quantum Big Bang Cosmology: Complex Space-time General Relativity, Quantum Coordinates, Dodecahedral Universe, Inflation, and New Spin 0, ½, 1 & 2 Tachyons & Imagyons™ (Pingree-Hill Publishing, Auburn, NH, 2004)

SuperCivilizations: Civilizations as Superorganisms (McMann-Fisher Publishing, Auburn, NH, 2010)

Standard Model Symmetries, And Four And Sixteen Dimension Complex Relativity; The Origin Of Higgs Mass Terms (Blaha Reasearch, Auburn, NH, 2012)

Available on bn.com, Amazon.com, Amazon.co.uk and other international web sites as well as at better bookstores (through Ingram Distributors).

Preface

The current major problems in particle physics and astrophysics are the source of possible new features: of The Standard Model, of the nature of Dark Matter, of the nature of Dark Energy, and of the nature of the Big Bang. In this book we propose an extended version of The Standard Model based on earlier work that adds another SU(2)⊗U(1) symmetry to the usual Standard Model, and an accompanying set of particles that we propose are the constituents of Dark Matter. This additional symmetry follows directly from a geometrical foundation for space-time within a 16 dimensional flat space that we call the Flatverse. Upon introducing a form of quantum coordinates we find that The Standard Model in those coordinates has no infinities, that the Big Bang is finite (no singularity), and that the Dark Energy that fuels the expansion of the universe has inflatons that consist of the imaginary part of the quantum coordinates – a free abelian gauge field. This field first stabilizes the universe in the Big Bang period and then causes a massive inflationary expansion. The complete theory has a remarkable convergence of features that remove infinities, identify the nature of Dark Matter and of Dark Energy, specify a physically acceptable Big Bang, and predict the observed expansion of the universe.

In addition we address the recently reported discovery of the Higgs particle that "explains" the origin of the masses of the other elementary particles but does not explain the origin of the Higgs particle mass terms and thus leaves open the ultimate question – What is the origin of mass and inertia? We show that the only final answer can be that it arises as a separation constant in Higgs dynamic equations that include coordinates of this universe and a sister universe. These two 8-dimensional universes are embedded in the 16-dimensional Flatverse. Thus the sister universe is the ultimate source of mass and inertia for our universe. The Flatverse is an absolute reference frame that is consistent with Einstein's General Relativity according to General Relativists. The Flatverse very nicely provides an environment for the two universes – joining them together to provide mass and inertia at their most fundamental level as well as providing, at last, a concrete definition of inertial reference frames. Result: a fundamental synergy between mass, inertia, and inertial frames.

In recent weeks major experimental findings on Dark Matter have been presented that at last begin to clarify the nature of Dark Matter and its interactions with "normal" matter. The findings reveal:

1) For the first time, decisive evidence that Dark Matter interacts with normal matter through forces other than gravitation.
2) Evidence now exists for at least three types of Dark Matter particles and their anti-particles. Dark Matter particles may be their own anti-particles (Majorana particles have this property.) But Dark particles may have different anti-particles.
3) The masses of Dark Matter particles appear to be about 8.6 GeV/c but have a range of values.
4) It appears that Dark Matter particles are electrically neutral but they can collide and produce electron-positron pairs. These facts suggest that a part of the known ElectroWeak interactions is also a Dark Matter interaction.
5) Dark Matter has its own family of forces in addition to a connection to the known ElectroWeak interactions. Dark Matter does not have the Strong Interactions.

The above findings, which remain to be fully confirmed, are consistent with a Dark Matter $SU(2)\otimes U(1)$ set of interactions that parallels, to a great extent, the known ElectroWeak interactions of normal matter. We will explore a detailed theory of Dark Matter interactions, and their relation to ElectroWeak interactions in this book. It is based on our previous work. In the previous work we did not specify a relation between ElectroWeak and "DarkWeak" interactions due to the absence of experimental data. With the information now in hand we can develop a detailed relationship.

Lastly, the book emphasizes again our belief that the origin of particle physics lies in a combination of Asynchronous Logic and space-time geometry.

CONTENTS

FIGURES and TABLES

1. Dark Matter Interactions with Normal Matter

The recent discoveries of new features of Dark Matter listed in the preface are consistent with our extended theory of The Standard Model of Elementary Particles presented below.[1] While many expected that very weak interactions would exist between normal matter and Dark Matter, earlier books by this author[2] did not introduce interactions in keeping with our conservative approach based on scrupulous consistency with known data about Dark Matter.

In this chapter we will propose a Weak interaction coupling between normal matter and Dark Matter based on the assumption of the known ElectroWeak interactions of normal matter and of a "parallel" SU(2)⊗U(1) set of "Dark ElectroWeak" Interactions for Dark Matter.

In earlier books we developed a derivation/construction of The Standard Model with eight possible forms due to lack of a theoretical or experimental basis to select a specific model. The derivation was based on a basic complex space-time geometry that was made real (as we experience it) through a group that we called the Reality group that mapped complex[3] space-time coordinates to real coordinates. *It led us in chapters 18 and 19 and Appendix 18-A of Blaha (2011) to derive an SU(3)⊗SU(2)⊗U(1)⊕U(1) Standard Model group representation based on the required form of the Reality group.*

Subsequently in Chapter 39 of Blaha(2012a) we saw that 4-dimensional complex coordinate space has a U(4) symmetry group with 16 generators which can be classified as the generators of the U(4) (non-commuting) subgroups: SU(3), SU(2), U(1), SU(2), and U(1). On that basis **we suggested that SU(3)⊗SU(2)⊗U(1)⊗SU(2)⊗U(1) was the complete Standard Model symmetry** with one SU(2)⊗U(1) symmetry being that of the known ElectroWeak

---

[1]Electron-Positron production through Weak Interaction with Dark Matter: M. Aguilar et al, Phys. Rev. Letters **110**, 141102 (2013). Three possible types of Dark Matter of mass 8.6 GeV/c (Three standard deviation effect but consistent with 2011 results of CoGeNT collaboration)): Talk at 13 April 2013 APS Meeting by team member Kevin McCarthy of the CDMS-II collaboration.

[2] Blaha (2011b), (2011c), (2012a) and (2012b).

[33] Complex space-time coordinates were required by the existence of superluminal neutrinos and quarks in the model. We shall describe the geometric basis of our Extended Standard Model in chapter 4.

interactions and the other SU(2)⊗U(1) symmetry being that of Dark Matter. Subsequent sections of Chapter 39 described three generations of Dark leptons and Dark SU(3) singlet quarks, and their overall chemistry and "Periodic Table."

Now that it appears that Dark Matter interacts weakly with normal matter, and that Dark Matter has a variety of particle types, we will now describe what appears to be the simplest interaction for Dark Matter and normal matter based on three working principles: 1) The only connecting interaction is a weak interaction, 2) The form of ElectroWeak theory remains unchanged, and 3) Dark Matter parallels normal matter in its general characteristics: three (possibly four) generations, an SU(2)⊗U(1) symmetry analogous to ElectroWeak symmetry, SU(2)⊗U(1) dark lepton and SU(3) singlet Dark quark doublets.

Chapter 2 describes the particle spectrum and chemistry of Dark Matter. The parallel to the known particle structure of normal matter will be evident. Dark Matter chemistry will be much simpler due to the absence of Strong interactions that result in nucleons and atomic nuclei as we see in normal matter. We conclude that Dark Matter life is not possible if Dark Matter leptons and quarks have Dark charges of equal magnitude.

1.1 The Standard Standard Model[4]

The lagrangian density for the conventional leptonic SU(2)⊗U(1) ElectroWeak Theory part of the Standard Model[5] with three or four generations is

$$\mathcal{L}_{EW} = \Psi_{3L}^{a\dagger}\gamma^0 i\gamma^\mu D_{L\mu}\Psi_{3L}^{a} - \Psi_{3R}^{a\dagger}\gamma^0 i\gamma^\mu D_{R\mu}\Psi_{3R}^{a} - \mathcal{L}_{BareMasses} + \mathcal{L}_{Gauge} + \mathcal{L}_{Mass} \tag{1.1}$$

where a is the generation index, and where

$$D_{L\mu} = D_\mu - \tfrac{1}{2} i g_2 \boldsymbol{\sigma}\cdot\mathbf{W}_\mu + i g_1 B_\mu/2 \tag{1.2}$$

$$D_{R\mu} = D_\mu + i g_1 B_\mu/2' \tag{1.3}$$

where D_ν is $\partial/\partial x^\nu$ and

$$\mathcal{L}_{Gauge} = -\tfrac{1}{4} F_W^{a\mu\nu} F_{W\,\mu\nu}^{a} - \tfrac{1}{4} F_B^{\mu\nu} F_{B\mu\nu} + \mathcal{L}_{EW}^{ghost} \tag{1.4}$$

---

[4] The repetition of the word "Standard" indicates the Standard Model of the past 35 years.

[5] Which we call The Standard Standard Model to distinguish it from our Extended Standard Model. Our alternate model we call The Extended Complexon Standard Model.

$\mathcal{L}_{BareMasses}$ contains the fermion bare mass terms if any. The ElectroWeak gauge bosons $W_\mu{}^a$, B_μ and B'_μ field tensors are:

$$F_{W}{}^{a}{}_{\mu\nu} = \partial W^a{}_\mu/\partial x^\nu - \partial W^a{}_\nu/\partial x^\mu + g_2 f^{abc} W^b{}_\mu W^c{}_\nu \quad (1.5)$$
$$F_{B_{\mu\nu}} = \partial B_\mu/\partial x^\nu - \partial B_\nu/\partial x^\mu \quad (1.6)$$

$\mathcal{L}_{EW}{}^{ghost}$ are the Faddeev-Popov ghost terms for the ElectroWeak $W_\mu{}^a$ gauge bosons.

The effective action for the path integral formulation is

$$I_{EW} = \int dx^0 d^3x\, \mathcal{L}_{EW} \quad (1.7)$$

We will not discuss the other sectors of this Standard Model at this point.

1.2 Extended Standard Models

In chapter 26 of Blaha (2011c)[6] we pointed out that due to theoretical and experimental uncertainty that there were eight possible varieties of Standard Model based on our geometric derivation:

Standard Model	**Number of Generations**	**Masses: Higgs (H) Dimensional (D)**	**Complexon Quarks/Gluons**
Standard Model 1	4	D	Yes
Standard Model 2	4	H	Yes
Standard Model 3	4	D	No
Standard Model 4	4	H	No
Standard Model 5	3	D	Yes
Standard Model 6	3	H	Yes
Standard Model 7	3	D	No
Standard Model 8	3	H	No

Figure 26.1. A Table of the possible variants of the "basic" Standard Model. Experiment is required to determine which of these models is the correct model.

---

[6] P. 316.

1.2.1 Types of Extended Models

The apparent discovery of Higgs bosons and the discovery of Dark Matter interactions, and their variety, reduces the number of possibilities to those listed in Fig. 1.1.

Standard Model	Number of Generations	Complexon Quarks/Gluons
Standard Model 2	4	Yes
Standard Model 4	4	No
Standard Model 6	3	Yes
Standard Model 8	3	No

Figure 1.1. Current table of the possible variants of the "basic" Standard Model. A Complexon quark is a quark with a complex 3-momentum. The real and imaginary parts of the 3-momentum are orthogonal with the result that the square of the momentum is purely real. These types of quarks are among the four possible fermion types described in Blaha (2011c) and earlier books.

The four types of Standard Models listed in Fig. 1.1 are for normal matter. The extension of the Standard Model possibilities to include Dark Matter interactions and symmetries greatly increases the number of possibilities. We shall consider one extension for Dark Matter below.

1.2.2 Extended Standard Model

There are several possible extensions of The Standard Model based on the selection of symmetry group for the Dark Matter sector. In the below sections we describe these possibilities. They are from Blaha (2011c) and chapters 38 and 39 of Blaha (2012a).

1.2.2.1 SU(3)⊗SU(2)⊕U(1) Standard Model

An extended Standard Model based on the symmetry group representation SU(3)⊗SU(2)⊗U(1)⊕U(1) with lepton and quark triplets (instead of the usual ElectroWeak doublets) is presented in appendix 18-A of Blaha (2011c). This model is not the physically correct one in the author's view based on the new experimental data on Dark Matter and the author's view that space-time geometry considerations favor an SU(3)⊗SU(2)⊗U(1)⊗SU(2)⊗U(1) Extended Standard Model as will be seen later.

1.2.2.2 SU(3)⊗SU(2)⊗U(1)⊗SU(2)⊗U(1) Standard Model Leptonic ElectroWeak Sector

This form of the extended Standard Model is simply defined with a lagrangian that uses a quadruplet of leptons – a pair of doublets instead of a single doublet. We will define a Dark matter sector with interaction with the normal matter sector in a minimal way.[7]

We define the left and right quadruplets respectively with

$$\Psi_{L,R}(x) = \begin{bmatrix} \psi_{DL,R}(x) \\ \psi_{NL,R}(x) \end{bmatrix} \tag{1.8}$$

where $\psi_{NL,R}(x)$ is a "normal" ElectroWeak-like lepton doublet, and where $\psi_{DL,R}(x)$ is a Dark ElectroWeak-like lepton doublet consisting of a Dark electron-like fermion and a Dark neutrino-like fermion.

We then define a lagrangian density

$$\mathscr{L}_{DN} = \overline{\Psi}_L(x) iD_L(x)\Psi_L(x) + \overline{\Psi}_R(x) iD_R(x)\Psi_R(x) + \mathscr{L}_{DNrest} \tag{1.9}$$

where $\mathscr{L}_{DNrest}$ contains the remaining gauge and Higgs terms of $\mathscr{L}_{DN}$. The covariant derivative terms are contained in D(x) which we express in matrix form as

$$D_{L,R}(x) = \begin{bmatrix} \gamma^\mu D_{DL,R\mu} & 0 \\ 0 & \gamma^\mu D_{NL,R\mu} \end{bmatrix} \tag{1.10}$$

where the normal matter left-handed covariant derivative is

---

[7] Based on the three working principles: 1) The only connecting interaction is a weak interaction, 2) The form of ElectroWeak theory remains unchanged, and 3) Dark Matter parallels normal matter in its general characteristics: three (possibly four) generations, SU(3) singlets, an SU(2)⊗U(1) symmetry analogous to ElectroWeak symmetry, SU(2)⊗U(1) dark lepton and dark quark doublets.

$$D_{NL\mu} = \partial/\partial x^\mu - \tfrac{1}{2}ig\boldsymbol{\sigma}\cdot\mathbf{W}_\mu + \tfrac{1}{2}ig'B_\mu \qquad (1.11)$$

and where the Dark matter left-handed covariant derivative is

$$D_{DL\mu} = \partial/\partial x^\mu - \tfrac{1}{2}ig_D\boldsymbol{\sigma}\cdot\mathbf{W}'_\mu + \tfrac{1}{2}ig_D'B'_\mu + \tfrac{1}{2}ig_D''B_\mu \qquad (1.12)$$

with $\boldsymbol{\sigma}$ a vector composed of the Pauli matrices. The right-handed covariant derivatives have a simpler form. The normal matter right-handed covariant derivative is

$$D_{NR\mu} = \partial/\partial x^\mu + ig'B_\mu \qquad (1.13)$$

and the Dark matter right-handed covariant derivative is

$$D_{DR\mu} = \partial/\partial x^\mu + ig_D'B'_\mu + ig_D''B_\mu \qquad (1.14)$$

Following the standard procedure for the normal matter sector we define the electromagnetic field A and the Z field by

$$\begin{aligned} W_3^\mu &= Z^\mu \cos\theta_W + A^\mu \sin\theta_W \\ B^\mu &= - Z^\mu \sin\theta_W + A^\mu \cos\theta_W \end{aligned} \qquad (1.15)$$

where θ_W is the Weinberg angle.

For the Dark Matter sector we define a Dark electromagnetic field A_D and a Dark Z_D field with analogous expressions based on the desire to have a similar overall formulation:

$$\begin{aligned} W'_3{}^\mu &= Z_D{}^\mu \cos\theta_D + A_D{}^\mu \sin\theta_D \\ B'^\mu &= - Z_D{}^\mu \sin\theta_D + A_D{}^\mu \cos\theta_D \end{aligned} \qquad (1.16)$$

where θ_D is an angle analogous to the Weinberg angle.

We now have to decide to what degree we can push the analogy between normal matter ElectroWeak interactions and Dark Matter "ElectroWeak" interactions. We will assume a strong analogy based on Nature's tendency to replicate successful structures (as seen in many areas of Physics and Biology). So we will assume a yet undiscovered Dark Higgs sector that gives mass to Dark Matter and Dark gauge boson fields but does *not* give mass to the Dark electromagnetic field A_D^μ. Thus Dark electric charge Q_D is conserved.

Eq. 1.12 can be rewritten in terms of Z_D^μ, A_D^μ and, most importantly, B_μ. The presence of B_μ in the Dark Matter covariant derivative enables the Dark covariant derivative term

$$D_{DL\mu} = \partial/\partial x^\mu - ½ig_D\boldsymbol{\sigma}\cdot\mathbf{W}'_\mu + ½ig_D'B'_\mu + ½ig_D''(-Z^\mu \sin\theta_W + A^\mu \cos\theta_W) \qquad (1.17)$$

to generate electron positron pairs as seen in the experiment of M. Aguilar et al. The reaction consists of Dark Fermion - Dark antifermion annihilation[8] into a virtual photon via the covariant derivative term $½ig_D''A^\mu\cos\theta_W$ in eq. 1.17 which then decays into an electron-positron pair:

$$D\bar{D} \rightarrow \gamma \rightarrow e^+e^- \qquad (1.18)$$

At the moment little is known about Dark Matter. But if the masses of Dark Matter particles is of the order of 8.6 GeV/c then the Dark Higgs sector must generate very large masses. Other points of interest within the framework of our assumptions are:

1. The gauge terms of the Dark Matter sector of the unified Lagrangian have the same form as those of the normal matter ElectroWeak sector.
2. Implicitly we have implemented the same Left-Right asymmetry as that of normal matter.
3. Dark Matter particles are SU(3) singlets. Dark Matter does not "have" the known Strong Interactions.
4. The Quark sector also has an SU(2)⊗U(1)⊗SU(2)⊗U(1) normal ElectroWeak Dark Matter "ElectroWeak" symmetry with the same general character.

---

[8] The possibility that Dark Matter is composed of Majorana Dark Particles whose anti-particles are the same as particles is not ruled out. It does, however, conflict with our notion that there is a parallel between normal matter and Dark Matter.

1.2.2.3 SU(3)⊗SU(2)⊗U(1)⊗SU(2)⊗U(1) Standard Model Quark ElectroWeak Sector

The quark ElectroWeak sector of the extended Standard Model is defined with a lagrangian that uses a quadruplet of quarks –a pair of doublets instead of a single doublet. We will define a quark Dark Matter sector that interacts with the normal matter sector in a minimal way.[9]

We define the left and right quark quadruplets with the same features as normal matter doublets. A Dark quark doublet is most naturally defined as consisting of a unit Dark charge quark and a neutral Dark charge quark. This assignment of Dark charge is based on the SU(3) singlet nature of Dark quarks and an analogy to the unit electric charge of physical normal particles such as hadrons.[10] (Dark quarks do not bind through the Strong interaction and so the Dark equivalent of baryons is single Dark quarks with unit Dark "electric" charge or zero charge.) However, Dark electromagnetism surrounds each Dark charged Dark quark with a cloud of Dark quarks and anti-quarks.

Turning to the Dark quark sector we define the left and right quark quadruplets respectively as

$$\psi_{qL,R}(x) = \begin{bmatrix} \psi_{DqL,R}(x) \\ \psi_{NqL,R}(x) \end{bmatrix} \tag{1.19}$$

where $\psi_{NqL,R}(x)$ is a "normal" ElectroWeak-like quark doublet, and where $\psi_{DqL,R}(x)$ is a Dark ElectroWeak-like quark doublet consisting of a Dark quark of unit Dark charge and a Dark quark of zero Dark charge.

---

[9] Again based on the three working principles: 1) The only connecting interaction is a weak interaction, 2) The form of ElectroWeak theory remains unchanged, and 3) Dark Matter parallels normal matter in its general characteristics: three (possibly four) generations, an SU(2)⊗U(1) symmetry analogous to ElectroWeak symmetry, SU(2)⊗U(1) Dark lepton and Dark SU(3) singlet quark doublets.

[1010] P. 189 of Sakurai (1964) makes the deep point, "One of the deepest mysteries in elementary particle physics centers around the question: Why is electric charge quantized? It is true that the present field theoretic formalism can explain why the positron charge is equal to the charge of the *physical* proton (in spite of the fact that the proton has a pion cloud) provided that the bare proton charge in the absence of the strong couplings is equal to the bare e^+ charge. However, we have as yet no compelling principle of minimal electromagnetic coupling to start with that requires that the bare p charge be equal to the bare e^+ charge." This same question applies to Dark charge although our field theoretic formalism also embodies Dark charge quantization.

We then define a quark sector lagrangian density

$$\mathscr{L}_{DN} = \bar{\Psi}_{qL}(x)iD_{qL}(x)\Psi_{qL}(x) + \bar{\Psi}_{qR}(x)iD_{qR}(x)\Psi_{qR}(x) + \mathscr{L}_{DNqrest} \quad (1.20)$$

where $\mathscr{L}_{DNqrest}$ contains the remaining gauge and Higgs terms. The covariant derivative terms are contained in $D_q(x)$ which we express in matrix form as

$$D_{qL,R}(x) = \begin{bmatrix} \gamma^\mu D_{qDL,R\mu} & 0 \\ 0 & \gamma^\mu D_{qNL,R\mu} \end{bmatrix} \quad (1.21)$$

where the normal quark matter left-handed covariant derivative is

$$D_{qNL\mu} = \partial/\partial x^\mu - \tfrac{1}{2} ig\boldsymbol{\sigma}\cdot\mathbf{W}_\mu - ig'B_\mu/6 \quad (1.22)$$

and where the Dark quark left-handed covariant derivative is

$$D_{qDL\mu} = \partial/\partial x^\mu - \tfrac{1}{2} ig_D\boldsymbol{\sigma}\cdot\mathbf{W}'_\mu + \tfrac{1}{2} ig_D'B'_\mu + \tfrac{1}{2} ig_D''B_\mu \quad (1.23)$$

since Dark quarks are SU(3) singlets with unit or zero Dark charge. The right-handed quark covariant derivatives have a simpler form. The normal quark right-handed covariant derivative is

$$D_{qNR\mu} = \partial/\partial x^\mu + ig'B_\mu/3 \quad (1.24)$$

and the Dark quark right-handed covariant derivative is

$$D_{qDR\mu} = \partial/\partial x^\mu + ig_D'B'_\mu + ig_D''B_\mu \quad (1.25)$$

Following the standard procedure for the normal quark sector we define the normal electromagnetic field A and the Z field by

$$\begin{aligned} W_3^\mu &= Z^\mu \cos\theta_W + A^\mu \sin\theta_W \\ B^\mu &= -Z^\mu \sin\theta_W + A^\mu \cos\theta_W \end{aligned} \quad (1.26)$$

where θ_W is the Weinberg angle.

For the Dark quark sector we can define the Dark electromagnetic field A_D and a Dark Z_D field with analogous expressions based on the desire to have a similar overall formulation:

$$W'^{\mu}_3 = Z_D^{\mu} \cos \theta_D + A_D^{\mu} \sin \theta_D \qquad (1.27)$$
$$B'^{\mu} = -Z_D^{\mu} \sin \theta_D + A_D^{\mu} \cos \theta_D$$

where θ_D is an angle analogous to the Weinberg angle.

Again we use the analogy between normal matter ElectroWeak interactions and Dark Matter ElectroWeak interactions. So we assume a Dark Higgs sector that gives mass to quark Dark Matter but does *not* give mass to the Dark electromagnetic field A_D^{μ}. Thus Dark electric charge Q_D is conserved in the Dark quark sector.

Eq. 1.23 can be rewritten in terms of Z_D^{μ}, A_D^{μ} and B_{μ}. The presence of B_{μ} in the Dark Matter covariant derivative enables the Dark covariant derivative to have the form

$$D_{qDL\mu} = \partial/\partial x^{\mu} - ½ig_D\boldsymbol{\sigma}\cdot\mathbf{W}'_{\mu} + ½ig_D'B'_{\mu} + ½ig_D''(-Z^{\mu} \sin \theta_W + A^{\mu} \cos \theta_W) \qquad (1.28)$$

Its form enables us to generate electron positron pairs as seen in the experiment of M. Aguilar et al. The reaction consists of Dark quark - Dark antiquark annihilation into a virtual photon via the covariant derivative term $½ig_D''A^{\mu}\cos \theta_W$ in eq. 1.28, which then decays into an electron-positron pair:

$$D\bar{D} \rightarrow \gamma \rightarrow e^+e^- \qquad (1.29)$$

Thus we have a tentative model for Dark Matter that accounts for the presently known experimental features.

1.2.2.4 SU(3)⊗SU(2)⊗U(1)⊗SU(2)⊗U(1) Complexon Standard Model ElectroWeak Sectors

In Blaha (2011c) we developed a Complexon Standard Model theory[11] in which quarks had complex 3-momenta whose inner product was a real number, and thus whose 4-momenta squared was a real number. The ElectroWeak sector of that theory can be transformed into a combined normal and Dark Matter ElectroWeak sector using the formalism develop above in subsections 1.2.2.2 and 1.2.2.3 by 1) leaving the leptonic sector (subsection 1.2.2.2) unchanged and 2) by replacing the derivative $\partial_\mu = \partial/\partial x^\mu$ in subsection 1.2.2.3 by

$$D_0 = \partial/\partial x^0$$
$$D_k = \partial/\partial x_r^{\ k} + i\, \partial/\partial x_i^{\ k} \qquad (1.30)$$

for k = 1, 2, 3 where $x_r^{\ k}$ is the real part of the kth coordinate and $x_i^{\ k}$ is the imaginary part of the kth coordinate. The normal and Dark ElectroWeak fields in eqns. 1.22 and 1.23 are functions of complex $x = x_r^{\ k} + ix_i^{\ k}$. The Faddeev-Popov mechanism applicable for these types of fields is described in appendix 19-A of Blaha (2011c). We denote the components in eq. 1.30 by D_μ. Then the *complexon* quark Standard Model ElectroWeak Sector covariant derivatives in quadruplet matrix form are

$$D_{qL,R}(x) = \begin{bmatrix} \gamma^\mu D_{qDL,R\mu} & 0 \\ 0 & \gamma^\mu D_{qNL,R\mu} \end{bmatrix} \qquad (1.31)$$

where the normal quark matter left-handed covariant derivative is

$$D_{qNL\mu} = D_\mu - ½ig\boldsymbol{\sigma}\cdot\mathbf{W}_\mu - ig'B_\mu/6 \qquad (1.32)$$

and where the Dark quark left-handed covariant derivative is

$$D_{qDL\mu} = \partial/\partial x^\mu - ½ig_D\boldsymbol{\sigma}\cdot\mathbf{W}'_\mu + ½ig_D'B'_\mu + ½ig_D''B_\mu \qquad (1.33)$$

---

[11] The formalism presented in this chapter does not use the quantum coordinates used in chapter 6 (and Blaha (2011c)) where the complete *finite* Complexon Standard Model is presented.

since Dark quarks are SU(3) singlets with unit or zero Dark charge. The right-handed complexon quark covariant derivatives have a simpler form. The normal quark right-handed covariant derivative is

$$D_{qNR\mu} = D_\mu + ig'B_\mu/3 \quad (1.34)$$

and the Dark complexon quark right-handed covariant derivative is

$$D_{qDR\mu} = D_\mu + ig_D'B'_\mu + ig_D''B_\mu \quad (1.35)$$

following the standard procedure as in eqns. 1.26 and 1.27. This leads to

$$D_{qDL\mu} = D_\mu - ½ig_D\boldsymbol{\sigma}\cdot\mathbf{W}'_\mu + ½ig_D'B'_\mu + ½ig_D''(-Z^\mu \sin\theta_W + A^\mu \cos\theta_W) \quad (1.36)$$

This covariant derivative term leads to the production of electron-positron pairs as well. The remaining parts of the complexon Standard Model are described in chapter 23 of Blaha (2011). The addition of Dark quark Higgs terms is also required.

1.3 Experimental Agreement with our Extension of Standard Models

The use of the preceding extension of our Standard Models, described in Blaha (2011) and subsequent books, gives us a set of models that agree with our current knowledge of Dark Matter.

Additional experiments are needed to determine whether quarks have complex 3-momenta, and whether there is a fourth generation of particles. More importantly, further experimental data on the properties of Dark Matter are woefully needed.

2. Particle Spectrum and Basic Chemistry of Dark Matter, Dark Matter Bodies

In chapter 39 of Blaha (2012a) we described Dark Matter particles, Dark atoms, the Dark Periodic Table, and Dark basic chemistry – all based on an additional SU(2)⊗U(1) symmetry that more or less mirrors normal ElectroWeak symmetry with the major differences being 1) that Dark quarks are assumed to be SU(3) singlets as suggested by the known weakness of Dark Matter interactions with normal matter and 2) that *physical* hadron charges – both electric charge and Dark electric charge – are quantized with whole number values. (Dark quarks are the hadrons of this theory since they do not experience the Strong Interaction.) This chapter contains material from chapter 39 of Blaha (2012a) for completeness as well as some additional thoughts.

Another important issue is the effect of gravitation on Dark Matter—Can Dark matter aggregate under the force of gravity to form galactic clusters, galaxies, and smaller objects such as Dark suns and perhaps Dark planets? Sections 2.3 and 2.4 discuss these possibilities.

2.1 Fundamental Dark Particles

We now consider the Dark particles that are associated with our new SU(2)⊗U(1) symmetry. Since Dark Matter only interacts weakly with known matter, Dark particles must be color singlets. Thus there will be 12 (in the three generation case) or 16 (in the four generation case) Dark particles plus their antiparticles. Recent experiments suggest Dark particles have extremely large masses of the order of 8.6 GeV/c or larger.

In addition to Dark fermions there will be four SU(2)⊗U(1) Dark gauge bosons – also with large masses:

Dark Particle Gauge Bosons

U(1): W'_0 SU(2): W'_1 W'_2 W'_3

The Dark fermions, which appear in Dark SU(2)⊗U(1) doublets are:

Dark Fundamental Fermions
(assuming 3 generations)

Leptonic Dark Particles	Quark Dark Particles
e_D	u_D
ν_{eD}	d_D
μ_D	c_D
$\nu_{\mu D}$	s_D
τ_D	t_D
$\nu_{\tau D}$	b_D

plus their anti-particles. Dark quarks are color singlets with complex 3-momenta in the extended Complexon Standard Model. In the extended "normal" Standard Model Dark quarks are color singlets with real 3-momenta.

2.2 Dark Particle Chemistry

The chemistry of Dark Matter will be different from the chemistry of known matter due to the absence of the color interaction. Dark particles cannot combine through a strong interaction to form a hadron spectrum such as we see in normal matter. Thus all Dark particle atoms will be like hydrogen atoms, and consist of a Dark quark particle bound to a Dark lepton particle by the Dark electric force assuming the Dark charge is quantized and has equal integer absolute values for Dark quarks and leptons.

Suppose all Dark charged particles have charge $\pm 1e_D$ (e_D is the Dark unit of charge) and each Dark doublet has a Dark particle with unit Dark charge and a Dark neutral Dark particle. Then we would expect that (in the case of three generations) Dark Matter would consist of

1. Lepton-like fundamental particles: Three Dark charged and three Dark charge neutral particles and their anti-particles.

2. Quark-like fundamental particles: Three Dark charged and three Dark charge neutral particles and their anti-particles.

3. There would be a total of six neutral Dark quarks and leptons.

4. Atoms are composed of oppositely charged Dark particles of different types. There are 9 Dark "atoms" of the form leptoDark particle - quarkDark particle, e_Du_D, e_Dc_D, e_Dt_D, μ_Du_D, μ_Dc_D, μ_Dt_D, τ_Du_D, τ_Dc_D, τ_Dt_D plus their anti-matter equivalents. There are four "quasi-stable" particle-anti-particle combinations: two leptoDark - antileptoDark particle combinations, and two quarkDark - antiquarkDark particle combinations. (There is no attractive nuclear force.) All of these combinations are bound by the Dark electric force.

5. Simple molecules of the type of Fig. 2.1 below based on Dark dipole interactions, Dark van der Waals forces and other Dark electromagnetic interactions are possible.

After a sufficiently long time collisions would lead perhaps to the dominance of Dark particles and the "disappearance" of antiDark particles if the number of Dark particles is overwhelmingly dominant in a fashion similar to normal matter.[12] (The other possibility is not excluded.) The Dark Periodic Table is:

Periodic Table of Dark Particle Atoms
(After Dark anti-particles are Annihilated)

e_Du_D	μ_Du_D	τ_Du_D	d_D
e_Dc_D	μ_Dc_D	τ_Dc_D	s_D
e_Dt_D	μ_Dt_D	τ_Dt_D	b_D

plus their similar antiparticle atoms. Bound states are assumed bound into hydrogen-like atoms through a Dark electromagnetic force. The last column consists of Dark charge neutral quarks. Antiparticle atoms of these states might also exist or be created through the Dark ElectroWeak interactions. One could extend the table with a column of Dark neutrinos. The decays, and mixing

---

[12] The study of electron-positron production through Weak Interaction with Dark Matter by M. Aguilar et al, Phys. Rev. Letters **110**, 141102 (2013) does not seem to clarify this issue.

between generations, remains to be determined by the unknown Dark Higgs sector.

The periodic table that we constructed is based on analogy with the features of normal matter: quarks are much heavier than leptons and leptons revolve around the Dark quark nuclei. If this view is correct then one can conceive of a chemistry of Dark Matter with molecules bound by Dark electromagnetic forces. Pair bonding of Dark leptons would be possible and so one could conceive of a fairly complex Dark chemistry bounded by the fact that a Dark particle atom has only one Dark lepton. Thus there would only be 18 bound pairs of atoms[13] similar to H_2.

$$\begin{array}{ccc} & e_D & \\ u_D & & u_D \\ & e_D & \end{array}$$

Figure 2.1. Example of two Dark atoms binding in a manner similar to the binding of hydrogen molecules in H_2. Considering the possible combinations of quarks and leptons there are 27 bound varieties of this type. Chains of these molecules are also possible in principle.

Dark dipole effects could lead to the (weaker) binding of larger assemblages of Dark atoms. For example, if the masses of the leptonic and quark Dark particles are not too dissimilar, crystalline Dark Matter appears possible.

We thus arrive at a Dark Matter sector with much less variety than normal matter. It may preclude the existence of Dark life and possibly of Dark solids composed of Dark Matter. Solid Dark planets may not exist.

Thus our universe is overwhelmingly composed of Dark Matter yet normal matter has far greater variety including the possibility of life.

2.3 Gravitation and Dark Matter Bodies: Galactic Clusters, Galaxies

Dark Matter is thought to equal 85% of the total matter in the universe. It is known to form halos around galaxies and to form filaments between galaxies in galactic clusters. It appears to influence the orbits of stars in galaxies causing

---

[13] Plus anti-particle equivalents.

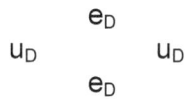

the rotational speed of stars at large radial distances from galactic centers to be roughly constant and independent of the radius.

Due to the relative lack of ElectroWeak and Strong interactions between Dark Matter and normal matter one can expect that gravitation will be the overwhelmingly major force between Dark Matter and normal matter.

We see this effect in the Dark Matter halos around galaxies. Galaxies are permeated with Dark Matter. Our galaxy, for example, appears to have ten times as much Dark Matter as normal matter.[14]

2.4 Dark Stars, Dark Planets, Dark Globular Clusters—Or Just Darkened?

However we have not seen (as yet?) clumps of Dark Matter within our galaxy similar in size to stars and planets or globular clusters. Nor do we see the earth, or other Solar System planets, having a discrepancy in mass due to Dark Matter clumps within these bodies.

Yet there is no apparent reason why Dark Matter should not clump within planets (and our sun) causing a discrepancy in solar system dynamics and ostensibly the force of gravity. The ten to one dominance of Dark Matter in the galaxy creates a striking discrepancy with the apparent absence of Dark Matter gravitational effects in the Solar System.

2.4.1 Open Questions on Dark Matter in Our Solar System

1. Why does not the earth's mass have a significant contribution from the Dark Matter within the earth since our galaxy has 10 times more Dark Matter than normal matter? Does not Dark Matter clump somewhat around planet size objects?

2. Same question as 1 but applied to the sun. Where is the effect of the Dark Matter within the sun?

3. Why is there no Dark Jupiter in our Solar System with normal moons circling it?

---

[14] Therefore one would expect more gravitational phenomena reflecting the presence of Dark Matter locally in our galaxy than its effect on galactic rotation. Since Dark Matter hardly interacts with normal matter it would intersperse with normal matter throughout the galaxy forming a hidden part of the galaxy – present but almost completely not detectable. The hidden part would occupy the same space as normal galactic matter rather like an evanescent ghost in a horror movie.

4. Why are not the planetary dynamics of the Solar System affected by the presence of Dark Matter?

2.4.2 Open Questions on Dark(ened) Stars and Dark(ened) Globular Clusters of Stars

There are many stars in the galaxy composed of normal matter apparently. However the ten to one ratio of Dark Matter to normal matter in our galaxy should be reflected in stellar dynamics which is a supposedly well understood field. Stellar dynamics depends significantly on the mass of stars.

1. Why does not stellar dynamics take account of the Dark Matter contribution to the mass of stars?
2. When a star contracts generating gravitational energy why does not the contraction of the Dark Matter clumped within the star impact gravitationally on the contraction of the normal matter?
3. Why is not the evolution and structure of globular clusters of stars affected by their presumably large Dark Matter content?
4. Will we find a Dark star circled by some large normal matter planets?
5. Can we detect a Dark cluster by its gravitational effects?

2.4.3 A Possible Solution to these Questions

The only currently apparent resolution of these questions is a lack of significant Dark chemical interactions/reactions between Dark Matter atoms that prevents the formation of Dark solids and liquids.[15] Dark nuclear decays similar to the normal nuclear reactions that power stars may also not exist. Thus no Dark stars.

Dark Matter in galaxies appears to be rather like a uniformly distributed gas within a gravitational well formed by the combined masses of a galaxy. It also can form a gravitationally bound gaseous filament between galaxies.[16]

Within a galaxy Dark Matter is dispersed in a more or less uniform way with only very minor gravitational clumping by clusters, stars and planets.

---

[15] Thus Dark atoms may not have Dark dipole forces or other multipole forces that would induce the formation of Dark Matter liquids and solids.

[16] A galaxy composed overwhelmingly of Dark Matter has recently been found.

3. Experimental Evidence for Faster-Than-Light Particles & Physics

Among the key assumptions of our extended Standard Models are 1) that the speed of light is the same in all inertial reference frames and 2) that some fundamental particles (neutrinos and down-type quarks) travel faster than the speed of light.

In this chapter we describe convincing evidence for faster than light physics. In chapter 4 we will show the need for a complex space-time that, suitably formulated, has a transformation between inertial reference frames that preserves the constancy of the speed of light in all inertial reference frames.

Until 1907 physicists thought that there was no limit on the speed of a particle or lump of matter. In 1907 Einstein and Poincaré showed that there was an inherent limit on the speed of a massive object – the speed of light. For the past 100 odd years physicists have generally accepted the speed of light as the limiting speed for particles with mass. Several theoretical physicists in the 1960's (E. C. Sudarshan and Gerald Feinberg) investigated the possibility of faster than light particles. They found that faster than light particles were theoretically possible but their theories – particularly their quantum field theories – had numerous discrepancies from canonical quantum field theory. These differences were taken by many to indicate that faster than light particles (called tachyons) were not present in nature. This belief was further supported by the happenings at particle accelerators where it was impossible to accelerate normal charged particles such as protons faster than the speed of light.

In the past ten years this author[17] developed a satisfactory quantum field theory of faster than light particles and found that if neutrinos and down-type quarks were faster than light particles he could derive the form of The Standard Model of Elementary Particles in detail. This theoretical development seems to have stimulated experimental groups at the new Linear Hadron Collider (LHC) at the CERN laboratory in Switzerland and the Gran Sasso Laboratory in Italy to

[17] See Blaha (2012b) and earlier books extending back nine years.

measure the speed of neutrinos emitted in LHC particle collisions. The results, described below, were mixed and one can fairly say they neither proved nor disproved that neutrinos were tachyons.

However there is other experimental data that strongly indicate that neutrinos are tachyons, and that quantum mechanics requires – not just faster than light behavior – but in some circumstances instantaneous effects at a distance – infinite speed of transmission!

In this chapter we will look at experimentally proven instantaneous Quantum Mechanical effects, at tritium decay experiments over the past 20 years that imply faster than light neutrinos, at neutrino speed measurements at the CERN LHC and Gran Sasso, at tachyonic particle behavior inside of Black Holes, and at the tachyonic behavior of Higgs particles, the "so-called God particle." *The cumulative result of these considerations is that faster than light particles, and physics, are a part of nature.*

3.1 Instantaneous Quantum Mechanical Effects

Quantum entanglement is a quantum phenomenon wherein parts of a physical system are in a certain quantum state but are separated by a space-like distance. If a change is made in part of a quantum entangled system then it is known theoretically, and experimentally, that other parts of the system change instantaneously.[18] Many experiments have shown that the change in other parts of a system is instantaneous and thus can be viewed as taking place at infinite speed – obviously beyond the speed of light.[19] The most recent experiment by Juan Yin et al[20] has shown directly that quantum mechanical effects travel faster than 10,000 times the speed of light. These experimental results are consistent with the instantaneous speed predicted by quantum mechanics. Thus faster than light behavior is implicit in quantum theory and is experimentally verified.

3.2 Tritium Decay Experiments Yielding Neutrinos

Fact: Particles with negative values for the square of their mass are tachyons – particles moving faster than light.

---

[18] Matson, John, "Quantum Teleportation Achieved Over Record Distances" *Nature* **13**, August 2012.

[19] Francis, Matthew, "Quantum Entanglement Shows that Reality Can't be Local", *Ars Technica*, 30 October 2012.

[20] Juan Yin et al, arXiv[quant-ph]: 1303.0614V1 (March 4, 2013).

A series of experiments by various groups over recent years imply that electron neutrinos produced in tritium decay have negative mass squared despite the best efforts of experimenters to obtain positive values for the neutrino mass squared.

Experiment	measured mass squared	Year
Mainz	$-1.6 \pm 2.5 \pm 2.1$	2000
Troitsk	$-1.0 \pm 3.0 \pm 2.1$	2000
Zürich	$-24 \pm 48 \pm 61$	1992
Tokyo INS	$-65 \pm 85 \pm 65$	1991
Los Alamos	$-147 \pm 68 \pm 41$	1991
Livermore	$-130 \pm 20 \pm 15$	1995
China	$-31 \pm 75 \pm 48$	1995
1998 Average	-27 ± 20	1998

Table 3.1 Electron neutrino mass squared values found in various tritium decay experiments. (Masses are in units of eV.) The average mass squared is negative suggesting electron neutrinos are tachyons.

Table 3.1 summarizes the measured electron mass squared in these experiments. These experiments strongly suggest that neutrinos have negative mass squared and are thus faster-than-light particles - tachyons.

3.3 LHC/Gran Sasso Direct Measurements of Neutrino Speeds

Two groups performed experiments at Gran Sasso Laboratory in Italy. They detected neutrinos emitted in interactions at the CERN LHC in Switzerland. The LVD collaboration in an exhaustive study of neutrino velocities found that the question was still open according to their data. Their refereed Physical Review Letter Abstract stated:

> We report the measurement of the time of flight of ~17 GeV ν_μ on the CNGS baseline (732 km) with the Large Volume Detector (LVD) at the Gran Sasso Laboratory. The CERN-SPS accelerator has been operated from May 10th to May 24th 2012, with a tightly bunched-beam structure to allow the velocity of neutrinos to be accurately measured on

> an event-by-event basis. LVD has detected 48 neutrino events, associated with the beam, with a high absolute time accuracy. These events allow us to establish the following limit on the difference between the neutrino speed and the light velocity: $-3.8\times10^{-6} < (v_\nu - c)/c < 3.1\times10^{-6}$ (at 99% C.L.). This value is an order of magnitude lower than previous direct measurements.[21]

These results (involving at least 35 neutrino detections) slightly favor, and do not rule out, faster-than-light neutrinos. Another experiment at the same locations by the ATLAS group stated that they found neutrino velocities (Five neutrinos were measured.) were below c. This group has not published their results as yet. We conclude that the published data appears to support faster than light neutrinos – consistent with our theory of The Standard Model.

A new project is in the planning stages to measure neutrino beams at larger distances. The hope is that the masses of the various neutrinos will be determined by the experiment. If the neutrino mass squared values turn out to be negative then it will constitute additional proof that neutrinos are tachyons (confirming tritium decay data), and thus support this author's formulation of The Standard Model of Elementary Particles.

3.4 Tachyonic Behavior Within Black Holes

Inside a black hole (such as the Schwarzschild solution of General Relativity) the time coordinate effectively becomes a spatial coordinate and the radius coordinate effectively becomes a time coordinate. An in-falling particle has a constantly decreasing radial distance from the center of the black hole just as time always increases outside a black hole.

As a result of the interchange of the roles of time and radius the velocity of a particle descending radially inside a Black Hole has a speed faster than light and is tachyonic.

3.5 Higgs Fields are Tachyons

Recently groups at the LHC CERN laboratory have announced the discovery of Higgs particles. The dynamic equations for Higgs bosons in The Standard Model have a negative mass squared. The mass squared must be negative or the Higgs Mechanism could not generate particle masses. Having

[21] N. Yu. Agafonova et al. (LVD Collaboration), "Measurement of the Velocity of Neutrinos from the CNGS Beam with the Large Volume Detector" Phys. Rev. Lett. **109**, 070801 (15 August 2012).

negative mass terms implies that Higgs fields are tachyonic – faster than light particles. Their tachyonic nature is masked by a quartic self-interaction that generates a condensate and thereby the masses of other particles.

3.6 Conclusion: Faster-Than-Light Particles – Tachyons Exist in Nature

The bulk of the experimental and theoretical evidence presented in previous sections strongly favors the existence of faster-than-light particles such as neutrinos. Tachyonic neutrinos are an important part of our form of The Standard Model. This form of the theory also strongly suggests that quarks are tachyonic in parallel with tachyonic neutrinos in order to obtain the symmetries of The Standard Model.

4. Geometric Basis of the Extended Standard Models

In a series of books[22] we have developed a new approach to the renormalization of quantum field theories, the first satisfactory theory of free tachyons, and a set of possible Standard Models that are derivable from fundamental space-time considerations combined with asynchronous logic. The set of possible Standard Models has been reduced by new experimental data and particularly by relating Dark Matter to a new ElectroWeak-like symmetry originating in space-time geometry.

Appendix A describes our derivation of the need for a 4-dimensional space-time based on the requirement that physical processes extend in both time and space necessitating a fundamental Asynchronous Logic basis for spatial and time synchronization. Asynchronous Logic can be implemented by a 4-valued logic which is represented by a 4-dimensional matrix formalism. This choice of fundamental basis eliminates the need for constructs such as strings which raise the further question: Where do they come from? In addition we note that Logic is an absolute necessity in the formulation of any physical theory: Without Logic, nothing! Thus it is the only thing that physics can hang on to with certainty.

In this chapter and in subsequent chapters we will outline the derivation of our forms of Standard Models. Most of it appears in our earlier books. But some significant new points, and a desire to present a rather complex theory in a higher level, unified way, has led us to present the discussion in this, and the following, chapters.

4.1 Rationale for Complex Coordinates and the Complex Lorentz Group

We assume a four dimensional physical space-time based on its derivation in Appendix A. As we will see, a reference frame moving faster than the speed of light relative to a "lab" reference frame will appear to have complex

---

[22] Blaha (2004) – Blaha (2012b). Parts of our development have changed. Blaha (2011c) – Blaha (2012b) have the correct theory with the addition of the Dark Matter discussions of this volume.

valued coordinates. Therefore we provisionally assume the 4-dimensional coordinate space is fundamentally complex.

We begin by assuming the speed of light is a common constant value in all inertial reference frames (Einstein). We also assume that all inertial reference frames are related by linear transformations so that an observer in one reference frame F can determine the coordinates of an event in another reference frame F′ by a linear transformation that preserves the value of the speed of light.

We then assume that the invariant distance between two infinitesimally separated events in an inertial frame has the form:

$$ds^2 = g_{\mu\nu}dx^{\mu}dx^{\nu *} \tag{4.1}$$

where $dx^{\nu *}$ is the complex conjugate of dx^{ν}. Eq. 4.1 implies that

$$g_{\mu\nu}{}^{*} = g_{\nu\mu} \tag{4.1a}$$

If we define a homogeneous transformation between F and F′ such that the invariant distance is preserved

$$ds'^2 = g_{\mu\nu}{}'dx'^{\mu}dx'^{\nu *} = ds^2 \tag{4.2}$$

then the form of the transformation must be a complex Lorentz transformation to satisfy eq. 4.2 and preserve the constancy of the speed of light in all inertial frames. If we denote the transformation as a 4 by 4 matrix Λ then it must satisfy

$$\Lambda^{T}G\Lambda = G \tag{4.3}$$

where Λ^T is the transpose of Λ and G is the 4 by 4 matrix of metric tensor components $g_{\mu\nu}$.

It will suffice for the present to consider the flat space-time case of a boost transformation from an "unprimed" coordinate system to a "primed" coordinate system moving with velocity v in the positive x direction. In matrix form we define an "unprimed" coordinate column vector with

$$a = \begin{bmatrix} t \\ x \\ y \\ z \end{bmatrix} \tag{4.4}$$

and the "primed" coordinates with

$$a' = \begin{bmatrix} t' \\ x' \\ y' \\ z' \end{bmatrix} \tag{4.5}$$

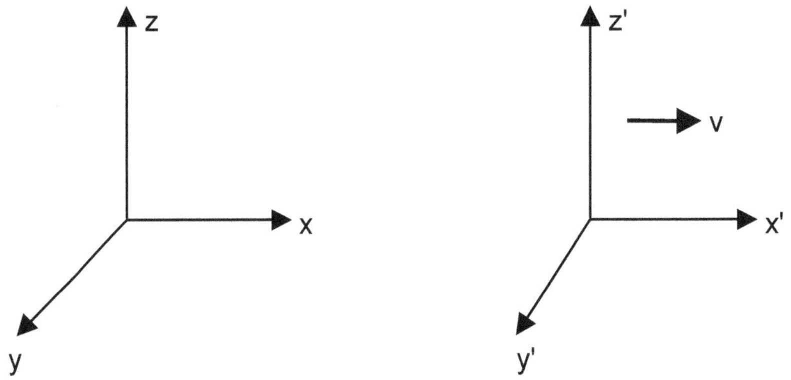

Figure 4.1. Depiction of two coordinate systems. The "primed" coordinate system is moving with velocity v in the positive x direction with respect to the "unprimed" coordinate system. We choose parallel axes for convenience.

The coordinates a and a' in this special case are related by a boost transformation Λ(v) with the form

$$a' = \Lambda(\mathbf{v})a \tag{4.6}$$

For eq. 4.2 to be valid the form of the boost transformation $\Lambda(\mathbf{v})$ is required to be[23]

$$\Lambda(\mathbf{v}) = \begin{bmatrix} \gamma & -\gamma v_x & -\gamma v_y & -\gamma v_z \\ -\gamma v_x & 1 + (\gamma - 1)v_x^2/v^2 & (\gamma - 1)v_x v_y/v^2 & (\gamma - 1)v_x v_z/v^2 \\ -\gamma v_y & (\gamma - 1)v_x v_y/v^2 & 1 + (\gamma - 1)v_y^2/v^2 & (\gamma - 1)v_y v_z/v^2 \\ -\gamma v_z & (\gamma - 1)v_x v_z/v^2 & (\gamma - 1)v_y v_z/v^2 & 1 + (\gamma - 1)v_z^2/v^2 \end{bmatrix} \quad (4.7)$$

where $\gamma = (1 - v^2)^{-½}$, $\mathbf{v} = (v_x, v_y, v_z)$, $v = |\mathbf{v}|$ and where we set $c = 1$ for convenience.[24] If $v < 1$ then $\Lambda(\mathbf{v})$ is real.

If $v > 1$, the superluminal case,[25] then $\Lambda(\mathbf{v})$ is complex and we find the a' coordinates are complex as well. Since faster than light (superluminal) motion is possible by chapter 3, we must accept transformations between observers in reference frames moving at a relative velocities faster than light. Therefore a fundamental *complex* 4-dimensional space-time is required. Faster than light transformations are required by the physical need to specify an event with coordinates in both the faster than light primed and the unprimed reference frames.

4.2 Complex Coordinates and the Reality Group

The fundamental complex coordinates of the primed reference frame found in the example in section 4.1 are a potential problem. We must understand the physical meaning of complex coordinates.

Realizing that an observer in any inertial coordinate system can *only* measure real-valued times and real-valued distances, we see that a further transformation is needed to transform a set of four complex coordinates to real

---

[23] We shall consider only the proper, orthochronous Lorentz group at this point. We assume that the primed and unprimed coordinate systems have parallel axes. So there is no rotation of axes embodied in eq. 4.6.

[24] One can set c = 1 by an appropriate choice of time and spatial distance scales. The demonstration that $\Lambda(\mathbf{v})$ has the form given by eq. 4.7 can be found in many textbooks.

[25] Note the form of $\Lambda(\mathbf{v})$ for $v > 1$ is the same as for $v < 1$. This commonness of form is required to preserve the constancy of the speed of light for $v > 1$ as well as $v < 1$.

values for any faster than light transformation. This additional transformation must be a U(4) group transformation.

Furthermore, since observers can accelerate with respect to each other,[26] causing position-dependent variation in the complexity of coordinates, the U(4) transformations must be local in the sense of Yang-Mills fields[27] in the case of accelerating reference frames.

In section 16.5 of Blaha (2011c) (and earlier work) we showed that U(4) has the subgroups SU(3), SU(2)⊗U(1), and another SU(2)⊗U(1) subgroup. These subgroups do not commute so they cannot be used directly to construct the SU(3)⊗SU(2)⊗U(1) group of the conventional Standard Model or that of the extended Standard Model of chapter 1: SU(3)⊗SU(2)⊗U(1)⊗SU(2)⊗U(1).

Nevertheless we can use U(4) which contains these subgroups as what we call the *Reality group*. It guarantees a transformation between reference frames always yields real coordinates. Thus we define the generalized inertial frame transformation for any real velocity (excepting $v = c$) by

v < c

$$\Lambda_T(\mathbf{v}) = I\Lambda(\mathbf{v}) \tag{4.8}$$

v > c

$$\Lambda_T(\mathbf{v}, \mathbf{x}) = \Lambda_R(\mathbf{v}, \mathbf{x})\Lambda(\mathbf{v}) \tag{4.9}$$

where I is the identity matrix and where $\Lambda_R(\mathbf{v}, x)$ is, in general, a local U(4) transformation. If the reference frames are not accelerating with respect to each other, then $\Lambda_R(\mathbf{v}, x)$ is independent of the coordinates **x**. Note for velocities $v < c$, $\Lambda_R(\mathbf{v}, \mathbf{x}) = I$ as eq. 4.8 shows and we have the usual Lorentz transformation.

Thus for $v > c$ we have an augmented transformation

$$a' = \Lambda_R(\mathbf{v}, \mathbf{x})\Lambda(\mathbf{v})a \tag{4.6a}$$

with the result that if the coordinates a are real-valued then the coordinates a' are also real-valued.

---

[26] Within the framework of flat space-times the transformation between accelerating coordinate systems is taken to be instantaneous Lorentz transformations plus their generalization to faster than light velocities.

[27] See Blaha (2011c) for a more detailed discussion.

4.3 Embedding Curved Space-time in the Flatverse

The U(4) Reality group has 16 generators. In four dimensions it is not possible to have separate commuting U(4) subgroups (SU(3), SU(2)⊗U(1), and another SU(2)⊗U(1)). The idea of using these subgroups as the origin of the Standard Model symmetries, while attractive, does not work because the group symmetry, R = SU(3)⊗SU(2)⊗U(1)⊗SU(2)⊗U(1), of the extended Standard Model has commuting factors.[28]

The simplest representation of the factors of R in which the factors are commuting groups and which can have a fully reducible representation is the 16 dimensional representation of U(16). Its 16 dimensional complex space supports a fully reducible R representation.

The use of a 16-dimensional complex space is interesting for another reason. Eq. 4.1a shows that the 4-dimensional metric tensor $g_{\mu\nu}$ has 16 independent components and so if one wishes to embed our complex 4-dimensional curved space-time within a flat space it must be at least be a 16-dimensional real space specified by sixteen equations:

$$z_i = f_i(x) \qquad (4.10)$$

where x is a complex 4-vector in our universe[29] (as in chapters 2 and 3 of Blaha (2012b).) The functions f_i map our universe into the 16 dimensional flat space as a 4-dimensional surface. Consistency with our 4-dimensional space-time suggests that the 16 dimension flat space should be complex.

Thus the requirement that the factors of R commute, and the embedding condition for our universe in a flat space, lead to the same 16 dimensional flat space. In chapter 5 of Blaha (2012b) we called this space the *Flatverse.*

The metric tensor of our universe $g_{\mu\nu}$ can be defined in terms of flatverse coordinates by

$$g_{\mu\nu} = \partial f_i/\partial x^\mu (\partial f_i/\partial x^\nu)^* \qquad (4.11)$$

with an implied sum over the subscript i.

---

[28] Some physicists have proposed an ultimate unification of all symmetries at some high energy based on an extrapolation of low order perturbation calculations. However there is no evidence that these commuting group factors will then become non-commuting. The origin of the various groups that constitute the Reality group in 4 dimensions shows that each performs a different role as shown in appendix 18-A and chapter 19 of Blaha (2011c).

[29] Our universe is then a 4-surface within the 16-dimensional flat space.

The complex 16 dimension Flatverse can also have a Reality group defined for it. After all if there were multiple universes between which travel was possible, then measuring distances again would necessarily be real-valued. Naturally it would be more elegant if the Reality group were G_{z16} = R = SU(3)⊗SU(2)⊗U(1)⊗SU(2)⊗U(1).

One can easily show that the 16 dimension Reality group must have 16 generators. G_{z16} has 16 generators. The other reasonable possibilities are SU(3)⊗SU(3) and U(4). U(4) is unacceptable because its generators do not have appropriate commutation relations – they would mix the strong and ElectroWeak interactions as will be apparent later. SU(3)⊗SU(3) has 16 generators—but the commutation relations of this group does not conform to the physical roles of the generators of the Reality group for 4 dimensional transformations. (See appendix 18-A and chapter 19 of Blaha (2011c).) Thus G_{z16} is the only reasonable choice of a 16 dimension Reality group.

The proof that G_{z16} is the correct choice for the 16 dimension Reality group follows from eq. 4.10. Suppose we perform a Reality group G_{z16} transformation on a 16 dimensional vector in the flatverse. Then there must be a corresponding Reality group transformation G_{x4} of the coordinates of our universe. So we can write

$$G_{z16a}z = f(G_{x4b}x) \tag{4.12}$$

where G_{z16a} is an element of G_{z16}, G_{x4b} is an element of G_{x4}, where z is a 16-dimensional vector and f is a vector composed of the 16 f_i functions. The 16 dimension group G_{z16} has 16 generators that we will denote V_i. The 4 dimension Reality group G_{x4} = U(4) has 16 generators that we will denote U_i. If we make an infinitesimal G_{z16a} transformation then

$$G_{z16a} = I + \alpha_i V_i \tag{4.13}$$

and G_{x4b} must be an infinitesimal transformation

$$G_{x4b} = I + \beta_i U_i \tag{4.14}$$

where α_i and β_i are constants[30] for I = 1, 2, …, 16. Substituting in eq. 4.12 and expanding to first order we find

$$\alpha_i V_{ijk} f_k = \beta_i U_i{}^{\mu}{}_{\nu}\, x^{\nu}\, \partial f_j/\partial x^{\mu} \tag{4.15}$$

for j = 1, 2, …, 16 with summations over i, k, μ, and ν. Eq. 4.15 provides 16 equations that determine the β_i parameters in terms of the α_i parameters. Thus a G_{z16} transformation uniquely determines the G_{x4} transformation.

In the case of a flat 4 dimensional space-time where we can limit the Flatverse to 4 dimensions also ($z_i = x_i$ for i = 1, 2, …,4), then the 16 dimension Reality group must be U(4) (not G_{z16}) and eq. 4.15 simplifies to[31]

$$\alpha_i V_{ijk} = \beta_i U_{ijk} \tag{4.16}$$

which immediately implies $\alpha_i = \beta_i$ and $V_{ijk} = U_{ijk}$. Thus the Flatverse's and our universe's Reality transformations coincide for a flat universe. In the case of a curved universe eq. 4.15 requires a more complex calculation. In any case, the equality of the number of Reality group generators in our universe and the Flatverse is crucial. Otherwise a solution is either ambiguous or non-existent in general.

4.4 Fully Reducible Representation of 16 Dimension Reality Group G_{z16}

The 16 dimension Reality group G_{z16} can be represented in a fully reduced form in terms of the regular representations of SU(3) – 8 by 8 matrices, SU(2) – 3 by 3 matrices, and U(1) 1 by 1 matrices. (See Fig. 2.2 of Blaha (2012b), which is reproduced here for the reader's convenience as Fig. 4.2.)

Denote the 16 matrix generators of G_{z16} as V_i. Corresponding to each generator is a connection or field that we will use to define covariant derivatives. We denote the connections as Z_{ki} where k labels the coordinate and i labels the connection. Then we divide these connections into sets of connections for each of the interactions that will ultimately become the extended Standard Model gauge fields:

---

[30] They could be functions of z and x respectively in the case of a curved 4-dimensional space-time or for accelerating reference frame transformations.

[31] We will not distinguish between raised and lowered indices for the sake of simplicity as they do not have physical import in these considerations.

SU(3) – A_{ik} for i = 1, 2, ..., 8 (4.17)
SU(2) – W_{ik} for i = 1, 2, 3
U(1) – W_{0k}
SU(2) – W'_{ik} for i = 1, 2, 3
U(1) – W'_{0k}

where k = 1, 2, …, 16 is the coordinate index.

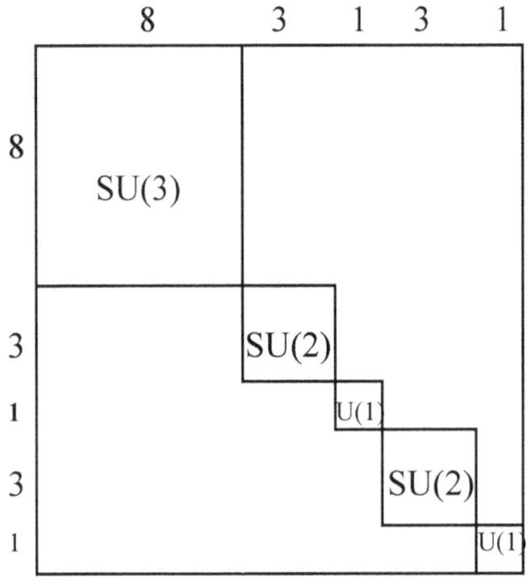

Figure 4.2. The fully reduced block structure of G_{z16} matrix representation.

Each matrix of each group's generators has its usual matrix regular representation in its block. It has 1's along the diagonal for all the other groups' blocks and zeroes otherwise. Thus each of the sets of generators has their conventional commutation relations although they are embedded in 16 by 16 matrices. For example the SU(3) generators would "fill" the top 8 by 8 block in Fig. 4.2. The rest of the generator matrices would consist of 1's along the diagonal following the block. All other components of its matrices would be zeros. See Fig. 4.3.

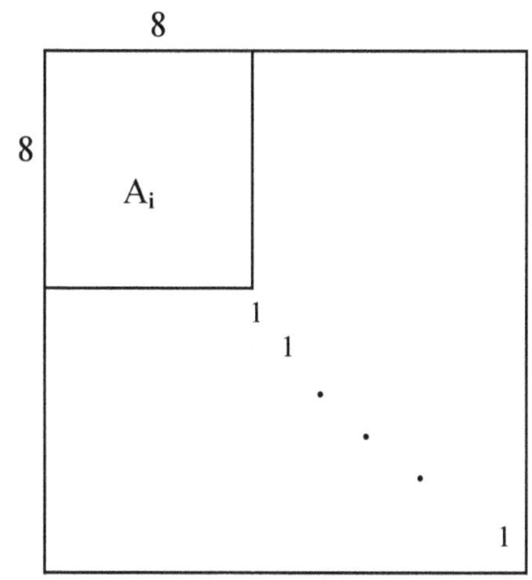

Figure 4.3. 16 dimension representation of the SU(3) generators.

4.5 Covariant Derivatives in the 16 Dimension Flatverse

The introduction of the Reality group in the Flatverse leads to the need for covariant derivatives for quantities that are subject to Reality group transformations. This question has been addressed in Blaha (2012a). In this section we will consider a simple example that introduces covariant derivatives necessitated by the Reality group.

Consider a 16 component vector function of the coordinates of the Flatverse **F**(z). If we apply a Reality group transformation G_{z16a} to it (to perhaps make all its components real-valued) then the partial derivative of G_{z16a}**F** changes in a non-covariant way.

$$\partial(G_{z16a}\mathbf{F})_j/\partial z^i \neq G_{z16ajk}\partial(\mathbf{F})_k/\partial z^i \tag{4.18}$$

In order to have a covariant derivative expression we must take the 16 generators of G_{z16}, which we denote V_i for i = 1, 2, ... , 16, and define the covariant derivative

$$D_k = \partial/\partial z^k - iV_iZ_{ki} \tag{4.19}$$

where there is an implicit sum over i. Using the covariant derivative we find

$$D_k(G_{z16a}\mathbf{F}) = G_{z16a}\partial\mathbf{F}/\partial z^k \tag{4.20}$$

4.6 Covariant Derivatives in Our 4 Dimension Universe

In the 16 dimension Flatverse we used 16 by 16 matrices to represent the Reality group taking advantage of the fully reduced structure of the Reality group's matrix representation. Interestingly the SU(3) and SU(2) blocks were 8 by 8 SU(3) regular representation matrices, and 2 by 2 SU(2) regular representation matrices.

In our 4 dimension universe we can use the 4 by 4 U(4) fundamental matrix representations for the Reality group. However the subgroups of U(4) do not commute making their ultimate use in particle symmetry groups impossible because the ElectroWeak, strong interaction, and the new SU(2)⊗U(1) group must commute.

Thus we must use R = SU(3)⊗SU(2)⊗U(1)⊗SU(2)⊗U(1), a direct product group, as the 4 dimension Reality group. The matrix representation of each factor group should be its fundamental representation, viz. 3 by 3, 2 by 2, 1 by 1, 2 by 2, and 1 by 1 respectively.[32]

The connections associated with the group R are listed in eq. 4.17 using the symbols normally used with Standard Model gauge fields. The connections are shown to become Standard Model gauge fields in Blaha (2011c) – particularly in appendix 18-A and chapter 19, and in chapter 1 earlier in this book.

In chapter 36 of Blaha (2012a) we developed a vierbein formulation of these considerations.

4.7 Extended Standard Model Gauge Fields

The connections listed in eq. 4.17 are parts of the covariant derivatives that appear in our extended Standard Model. Their use in the extended Standard Model is described in Blaha (2011c) and the extension of The Standard Model described in chapter 1. They describe the lagrangian terms, the Faddeev-Popov methods and other aspects of these gauge fields. Thus the gauge field sector of the extended Standard Model is fully described.

The other parts of the extended Standard Model: fundamental fermions and Higgs particles are also described in Blaha (2011c), (2012a) and (2012b).

[32] It is intriguing that the 16 dimension Reality group uses the regular representations of the group factors while the 4 dimension Reality group uses their fundamental representations. An additional point of interest is U(4)'s fundamental matrix representation consists of 4 by 4 matrices while its regular representation consists of 16 by 16 matrices. One sees a regularity of features that argues for a deeper cause as yet unknown.

Later in this book we will present these parts from a slightly different viewpoint which we hope will make it clearer to readers.

4.8 Quantizing Our Universe's Coordinates within the Flatverse

Eq, 4.10 prescribes the embedding of our universe as a surface within the 16 dimension Flatverse. These equations are c-number equations. In Blaha (2005a) we showed that the use of slightly q-number coordinates

$$X^{\mu}(z) = z^{\mu} + i\, Y^{\mu}(z)/M_c^{\,2} \qquad (4.21)$$

(where M_c is a large mass scale of perhaps the order of the Planck mass, and where $Y^{\mu}(y)$ is a free QED-like field) in any quantum field theory enabled us to eliminate infinities to any order of perturbation theory. We called this type of quantum field theory a Two-Tier quantum field theory.

In chapter 4 of Blaha (2012b) we showed that the Two-Tier formulation of a quantum field theory could be viewed as an embedding our universe in the Flatverse for the special case of a flat universe which would allow the Flatverse to be 4 dimensional. The generalization to a curved space-time embedded in a 16 dimension Flatverse is direct.

We also showed in chapter 5 of Blaha (2005a) that a particle field in z^{μ} coordinates can be "dressed" to be a particle field in $X^{\mu}(z)$ coordinates.

Thus one can view the Flatverse as the "source" of the bare particles which can be dressed to cloaked particles in our universe with q-number coordinates. The cloaking of particles surrounds them with a cloud of $Y^{\mu}(z)$ vector bosons that effectively smears the particles in such a way as to suppress infinities in perturbation calculations.[33]

4.9 Two-Tier Quantized Coordinates For Complexons

In chapter 6 we will consider complexon quarks – quarks with complex 3-momenta. In this case we must use a generalization of eq. 4.21 to complex z coordinates. In this section we will outline this generalization[34]

---

[33] See Blaha (2005a).

[34] See Blaha (2011c) section 19.5.2 and following for more detail.

In the case of the complexon quantum fields (quarks and color gauge particles) we will need two q-number variables X_r^μ and X_i^μ since we have complex spatial 3-coordinates. We define them similarly to the previous case:

$$X_{r_\mu}(y_r) = y_{r_\mu} + i\, Y_{r_\mu}(y_r)/M_c^2 \tag{4.22}$$

$$X_{i_\mu}(y_i) = y_{i_\mu} + i\, Y_{i_\mu}(y_i)/M_c^2 \tag{4.23}$$

$$X_{c_\mu}(y_i) = X_{r_\mu}(y_r) + i\, X_{i_\mu}(y_i) \tag{4.23a}$$

where we choose the same mass scale for both the "real" and "imaginary" variables.[35] The Two-Tier, single generation version of the Complexon Standard Model (chapter 6) then has an action of the form

$$I_{CSMtt} = \int dy^0 d^3y_r d^3y_i\, (\mathscr{L}_{CSM}(X_r^\mu(y_r), \mathbf{X}_i^k(y_i))J_2)\big|_{y_i^0 = 0,\ Y_r^0 = Y_i^0 = 0} +$$

$$+ \int dy_r^0 d^3y_r\, \mathscr{L}_C(X_r^\mu(y_r), \partial X_r^\mu(y_r)/\partial y_r^\nu, y_r) +$$

$$+ \int dy_i^0 d^3y_i \mathscr{L}_C(X_i^\mu(y_i), \partial X_i^\mu(y_i)/\partial y_i^\nu, y_i) \tag{4.24}$$

where the replacements

$$x^\mu \equiv x_r^\mu \rightarrow X_r^\mu(y_r) \tag{4.25}$$

$$x_i^k \rightarrow X_i^k(y_i) \tag{4.26}$$

for $\mu = 0, 1, 2, 3$ and $k = 1, 2, 3$ are made in $\mathscr{L}_{CSM}$ followed by defining $y_r^0 = y^0$ and making an L(C) transformation to a frame where $y_i^0 = 0$. J_2 is the absolute value of the Jacobian of the transformation from (X_r, X_i) to (y_r, y_i) coordinates:

---

[35] Note that both the real and imaginary coordinates have an imaginary q-number part. The q-number nature of these imaginary parts allows us to distinguish them from the c-number parts of the coordinates. This distinguishing feature is important in perturbative calculations as can be seen in examples in this book and earlier books.

$$J_2 = |\partial(X_r, X_i)/\partial(y_r, y_i)| \quad (4.27)$$

We also choose gauges where

$$Y_r^0 = Y_i^0 = 0 \quad (4.23b)$$

These types of transformations and gauge choices are discussed in detail in Blaha (2005a). The lagrangian terms $\mathscr{L}_c(X_r^\mu(y_r), \partial X_r^\mu(y_r)/\partial y_r^\nu, y_r)$ and $\mathscr{L}_c(X_i^\mu(y_i), \partial X_i^\mu(y_i)/\partial y_i^\nu, y_i)$ have the same form:

$$\mathscr{L}_C = +\tfrac{1}{4} M_c^4 F^{\mu\nu} F_{\mu\nu} \quad (4.28)$$

with

$$F_{\mu\nu} = \partial X_\mu/\partial y^\nu - \partial X_\nu/\partial y^\mu \quad (4.29)$$

$$\equiv i\,(\partial Y_\mu/\partial y^\nu - \partial Y_\nu/\partial y^\mu)/M_c^2 \quad (4.30)$$

or defining

$$F_{Y_{\mu\nu}} = (\partial Y_\mu/\partial y^\nu - \partial Y_\nu/\partial y^\mu) \quad (4.31)$$

we see each lagrangian assumes the form of the conventional electromagnetic Lagrangian:

$$\mathscr{L}_C = -\tfrac{1}{4} F_Y^{\mu\nu} F_{Y_{\mu\nu}} \quad (4.32)$$

The lagrangian is supplemented with the following condition on all complexon fields $\Phi_{...}$:[36]

$$(\partial/\partial X_r^k(y_r))\,(\partial/\partial X_i^k(y_i))\Phi... = 0 \quad (4.33)$$

summed over k = 1, 2, 3. Non-complexon fields Ω... in our left-handed formulation satisfy the subsidiary condition:

$$\{(\partial/\partial X_r^k(y_r))(\partial/\partial X_i^k(y_i)) - [(\partial/\partial X_r^k(y_r))^2(\partial/\partial X_i^m(y_i))^2]^{1/2}\}\Omega... = 0 \quad (4.34)$$

---

[36] Eq. 4.33 and 4.34 are required by the particle complex 3-momentum constraint eq. 5.50. This constraint ensures fundamental particle energies are real. See the discussion of the rationale in chapter 5.

summed over k = 1, 2, 3 and over m = 1, 2, 3 separately in each of the two terms.

The momentum space, free field, Feynman propagators G...(k) of all particles and ghosts in Two-Tier QFT acquire a Gaussian factor exp(h(k)):

$$G...(k) \rightarrow G...(k) \exp(h(k)) \tag{4.35}$$

that causes all perturbation theory diagrams to be finite. The consequence is finite results in all calculations to any order in perturbation theory. Blaha (2005a) shows that Two-Tier theories are finite, Poincaré covariant, and unitary.

An example of the Two-Tier effect on propagators is the Two-Tier photon propagator. The Two-Tier photon propagator[37] in Two-Tier QED is:

$$iD_F^{TT}(y_1 - y_2)_{\mu\nu} = -i \int \frac{d^4p\, e^{-ip\cdot z}\, g_{\mu\nu} R(\mathbf{p}, z)}{(2\pi)^4 (p^2 + i\varepsilon)} \tag{4.36}$$

(since the imaginary parts can be taken to be zero: $y_{1i}^{\mu} - y_{2i}^{\mu} = 0$) where

$$z^{\mu} = y_{1r}^{\mu} - y_{2r}^{\mu} \tag{4.37}$$

$$R(\mathbf{p}, \mathbf{z}) = \exp[-p^i p^j \Delta_{Tij}(z)/M_c^4] \tag{4.38}$$

$$= \exp\{-\mathbf{p}^2[A(v) + B(v)\cos^2\theta] / [4\pi^2 M_c^4 |\mathbf{z}|^2]\} \tag{4.39}$$

with i, j = 1, 2, 3, and with $\Delta_{Tij}(z)$ the commutator of the positive frequency part $Y^{+}_{k}(y)$ and the negative frequency part $Y^{-}_{k}(y)$ of $Y_k(y)$:

$$\Delta_{Tij}(z) = [Y^{+}_{j}(y_{1r}), Y^{-}_{k}(y_{2r})] = \int d^3k\, e^{ik\cdot(y_{1r} - y_{2r})} (\delta_{jk} - k_j k_k/\mathbf{k}^2)/[(2\pi)^3 2\omega_k] \tag{4.40}$$

and

$$v = |z^0|/|\mathbf{z}|$$

$$A(v) = (1 - v^2)^{-1} + .5v \ln[(v - 1)/(v + 1)] \tag{4.41}$$

$$B(v) = v^2(1 - v^2)^{-1} - 1.5v \ln[(v - 1)/(v + 1)] \tag{4.42}$$

$$\mathbf{p}\cdot\mathbf{z} = |\mathbf{p}|\ |\mathbf{z}| \cos\theta \tag{4.43}$$

---

[37] Blaha (2005a).

with |**p**| denoting the length of a spatial vector **p**, |**z**| denoting the length of a spatial vector **z**, and with $|z^0|$ being the absolute value of z^0.

The gaussian factors R(**p**, z) which appear in all Two-Tier propagators damp the large momentum behavior of all perturbation theory integrals producing a completely finite perturbation theory and *yet give the usual results of perturbation theory at energies that are small compared to the mass scale* M_C.

In the case of complexons, the Two-Tier Feynman propagator differs from the non-complexon case by having an integration over imaginary spatial 3-momenta, a derivative of a delta function embodying the orthogonality of the real and imaginary 3-momenta, and two factors of R(**p**, z): one factor being R($\mathbf{p}_r$, z_r) and the other factor being R($\mathbf{p}_i$, z_i) (where the time components $z_r^0 = z^0$ and $z_i^0 = 0$ since there is only one real time coordinate[38]). Thus we obtain large momentum convergence for both real and imaginary 3-momentum integrations.[39]

For a scalar complexon particle in conventional quantum field theory we find the Feynman propagator:

$$i\Delta_{CTF}(x-y) = \theta(x^+ - y^+)\langle 0|\phi_{CT}(x)\,\phi_{CT}(y)|0\rangle + \theta(y^+ - x^+)\langle 0|\phi_{CT}(y)\phi_{CT}(x)|0\rangle$$

$$= i\int \frac{d^4p_r d^3p_i(2\pi)^{-7}\delta'(\mathbf{p}_r\cdot\mathbf{p}_i/m^2)e^{-ip^+(x^- - y^-) - ip^-(x^+ - y^+) + i\mathbf{p}_\perp\cdot(\mathbf{x}_\perp - \mathbf{y}_\perp) - i\mathbf{p}_i\cdot(\mathbf{x}_i - \mathbf{y}_i)}}{(p^2 + m^2 + i\varepsilon)} \tag{4.44}$$

In the case of Two-Tier quantum field a scalar complexon has the the Feynman propagator

$$i\Delta_{CTFtt}(x-y) = i\int d^4p_r d^3p_i(2\pi)^{-7}\delta'(\mathbf{p}_r\cdot\mathbf{p}_i/m^2)\, R(\mathbf{p}_r, z_r)R(\mathbf{p}_i, z_i)\cdot$$
$$\cdot\, e^{-ip^+(x^- - y^-) - ip^-(x^+ - y^+) + i\mathbf{p}_\perp\cdot(\mathbf{x}_\perp - \mathbf{y}_\perp) - i\mathbf{p}_i\cdot(\mathbf{x}_i - \mathbf{y}_i)}/(p^2 - m^2 + i\varepsilon) \tag{4.45}$$

---

[38] We can arrange for $z_i^0 = 0$ by making a L_C transformation to an inertial frame where z is real.
[39] For a detailed discussion see Blaha (2011c).

where the time components $z_r^0 = z^0$ and $z_i^0 = 0$ since there is only one time coordinate and where $p^2 = p^{0\,2} - p_r^2 + p_i^2$. Note both the real and imaginary integrations have gaussian damping at high momenta.[40]

Propagators for other types of particles are similarly modified in the Two-Tier formalism (See Blaha 2005a).

The above discussion of Two-Tier theory appears to be quite complicated but it is simple enough in practice. One important feature of Two-Tier theory is that it produces the same results as conventional perturbation theory at low energies (E << M_c.) The mass scale M_c is very large – perhaps of the order of the Planck mass.

4.10 Dark Energy, The Big Bang and Quantized Coordinates

The existence of a Y particle cloud around all the particles in the universe, including gravitons, is a source of enormous energy. In addition to eliminating infinities in quantum field theory, the Y particles within the universe are free (no interactions!) and a major contribution to the stress-energy tensor that governs the structure of the universe. In chapter 10 we show how these particles prevent singularities at the beginning of the universe and provide the Dark Energy needed for the observed expansion of the universe. [41]

Thus we find a universe with three parts:

1. The known Standard Model particles.

2. The additional particles described in our extended Standard Model (chapters 1 and 2) that we believe constitute Dark Matter.

3. The Y particles appearing in quantized coordinates that constitute Dark Energy and are responsible for the overall evolution of the universe. (Chapter 10)

Thus we have an overall theory that accounts for all known features of the universe.

---

[40] Due to the imaginary q-number parts in eqs 4.22 and 4.23.

[41] See Blaha (2004) for a model of the Big Bang period and the subsequent expansion of the universe. This model is presented with some changes in chapter 10.

5. Fermion Spectrum

The fundamental fermions part of the Standard Models consist of a set of spin ½ particles with a clear pattern rather analogous to the Periodic Table of Chemistry. In chapters 17-19 and part 4 of Blaha (2011c), and in earlier work, we showed that there are fundamental reasons for the pattern of fermions stemming from the geometry of space-time that we have developed based on the complex Lorentz group and the Reality group.

In this chapter we will describe the origin of the form of the fermion spectrum in complex Lorentz group and Reality group features.[42]

5.1 The Fermion Spectrum

The currently known fermion spectrum is

Known Fundamental Fermions
(assuming 3 generations)

Leptonic Particles	Quark Particles
e	u_i
ν_e	d_i
μ	c_i
ν_μ	s_i
τ	t_i
ν_τ	b_i

where i = 1, 2, 3 labels the quark color. In chapter 1 we conjectured that given the SU(2)⊗U(1) proposed Dark symmetry that the Dark particles would have a similar form to the known fundamental fermions with one difference—Dark quarks would be SU(3) singlets—since they do not have the strong interaction. As a result we suggested the Dark fermion spectrum would be:

[42] Much of this chapter appears in a somewhat different form starting in chapter 19 of Blaha (2011c) and earlier books.

Dark Fundamental Fermions (assuming 3 generations)

Leptonic Dark Particles	Quark Dark Particles
e_D	u_D
ν_{eD}	d_D
μ_D	c_D
$\nu_{\mu D}$	s_D
τ_D	t_D
$\nu_{\tau D}$	b_D

where the subscript D indicates a Dark particle. The other names and subscripts follow the same pattern as normal fermions.

In this section we will propose a rationale for the spectrum of fermions based on the detailed study of fermion characteristics in Blaha (2011c) and earlier books.

The first observation on fermions is that there appears to be four species of fermions that correspond to

1. Leptons – Two species - Dirac fermions with real valued momenta, tachyons with real valued momenta,
2. Quarks – Two species - Dirac-like fermions with complex spatial 3-momenta, and tachyon-like fermions with complex spatial 3-momenta.

Secondly, Dirac and tachyon-like fermions with real valued momenta are grouped into lepton ElectroWeak doublets for normal fermions, and grouped into Dark lepton SU(2)⊗U(1) doublets for Dark fermions.

Thirdly, Dirac and tachyon-like fermions with complex valued momenta are grouped into quark ElectroWeak doublets for normal fermions, and grouped into Dark quark SU(2)⊗U(1) doublets for Dark fermions.

Fourthly, normal quarks are color SU(3) triplets and Dark quarks are color SU(3) singlets since they do not exhibit the strong interaction.[43]

---

[43] We rule out a Dark strong interaction (color) because it would require an additional SU(3) factor in the Reality group – contrary to our prior discussions on the form of the Reality group which clearly does not have such a factor.

The above four observations follow directly from our space-time discussions of Blaha (2011c) and chapter 1 of this book. Thus we derive the form of the fermion spectrum from the complex Lorentz group and the Reality group.

The three known generations of fermions yield the below Periodic Table. Some preliminary evidence for a fourth generation has appeared. A four generation spectrum would be more natural within the framework of our approach.

"Periodic" Table of Fermions

NORMAL ELECTROWEAK FERMION DOUBLETS

Generation	ElectroWeak Leptons Real-Valued 3-Momenta		Color Triplet Quarks Complex-Valued 3-Momenta	
1	e	ν_e	u_i	d_i
2	μ	ν_μ	c_i	s_i
3	τ	ν_τ	t_i	b_i
?				

DARK SU(2)⊗U(1)FERMION DOUBLETS

Generation	Dark SU(2)⊗U(1) Leptons Real-Valued 3-Momenta		Color Singlet Quarks Complex-Valued 3-Momenta	
1	e_D	ν_{eD}	u_D	d_D
2	μ_D	$\nu_{\mu D}$	c_D	s_D
3	τ_D	$\nu_{\tau D}$	t_D	b_D

The next section outlines the derivation of the above spectrum. See chapters 17-19 and part 4 of Blaha (2011c) for a more detailed discussion.

5.2 Complex Space-Time Origin of the Four Fermion Species

The four types of fermion species that we have found in our theoretical analyses have distinct defining features. These fermion features – fermions are normal or tachyons, and fermions have real 3-momenta or complex 3-momenta – remain to be proven experimentally. Chapter 3 shows that there is good reason to believe neutrinos are tachyons proving one part of our theory. The theoretic proof that quarks have complex 3-momenta remains to be verified experimentally. One problem, that makes experimental proof of complex 3-momenta difficult, is that the only handle we have on quark behavior – parton models – can be created that work well enough generally with real 3-momenta because real and imaginary 3-momenta parts are separately conserved. One possible way of experimentally measuring imaginary 3-momenta parts of partons is through the impact of imaginary 3-momenta on quark parton spin dynamics. The parton spin structure functions disagree with experiment in significant ways. A reanalysis of the experimental data under the assumption that quark partons have imaginary 3-momenta parts which are perpendicular to their real 3-momenta parts would be informative and possibly resolve the discrepancy between parton spin structure functions in theory and experiment.

We now turn to showing the origin of the four species of fermions as developed in Blaha (2011c) and earlier books. We assume that every fermion has a (perhaps miniscule) bare mass so that it has a rest state. We then consider all possible L(C) boosts[44] to generate fermions with non-zero 4-momentum.

A fundamental requirement for fundamental free fermions is that they have a real-valued energy. If they did not, then they would decay and thus not be fundamental.

L(C) boosts have the form

$$\Lambda_C = \exp[i\mathbf{a}\cdot\mathbf{K}] \tag{5.1}$$
$$= \exp[i(\omega_r\hat{\mathbf{u}}_r + i\omega_i\hat{\mathbf{u}}_i)\cdot\mathbf{K}] \tag{5.2}$$

where K is the Lorentz boost generator and the 3-vector **a** is complex. Eq. 5.2 expresses **a** in terms of complex constants: real constants $\omega_r \geq 0$ and $\omega_i \geq 0$. The vectors $\hat{\mathbf{u}}_r$ and $\hat{\mathbf{u}}_i$ are real normalized 3-vectors such that $\hat{\mathbf{u}}_r\cdot\hat{\mathbf{u}}_r = 1 = \hat{\mathbf{u}}_i\cdot\hat{\mathbf{u}}_i$.

The general form of an L(C) boost can also be expressed in the form

$$\Lambda_C(\mathbf{v}_c) \equiv \Lambda_C(\omega, \mathbf{v}_c) = \exp[i\omega\hat{\mathbf{w}}\cdot\mathbf{K}] \tag{5.3}$$

where

$$\omega = (\omega_r^2 - \omega_i^2 + 2i\omega_r\omega_i\, \hat{\mathbf{u}}_r\cdot\hat{\mathbf{u}}_i)^{1/2} \tag{5.4}$$

and

$$\hat{\mathbf{w}} = (\omega_r\hat{\mathbf{u}}_r + i\omega_i\hat{\mathbf{u}}_i)/\omega \tag{5.5}$$

---

44 L(C) is the complex Lorentz group. We shall see that complex coordinates are required in the consideration of tachyon (faster than light) particles

If we limit[45] the L(C) boosts to the set of boosts $L_B(C)$ where $\hat{\mathbf{u}}_r \cdot \hat{\mathbf{u}}_i = 0$ then

$$\hat{\mathbf{w}} \cdot \hat{\mathbf{w}} = 1 \tag{5.6}$$

and the complex relative velocity is

$$\mathbf{v}_c = \hat{\mathbf{w}} \tanh(\omega) \tag{5.7}$$

The matrix form of proper (det $\Lambda_C = 1$) $L_B(C)$ coordinate boosts is

$$\Lambda_C(\mathbf{v}_c) = \begin{bmatrix} \cosh(\omega) & -\sinh(\omega)\hat{w}_x & -\sinh(\omega)\hat{w}_y & -\sinh(\omega)\hat{w}_z \\ -\sinh(\omega)\hat{w}_x & 1 + (\cosh(\omega) - 1)\hat{w}_x^2 & (\cosh(\omega) - 1)\hat{w}_x\hat{w}_y & (\cosh(\omega) - 1)\hat{w}_x\hat{w}_z \\ -\sinh(\omega)\hat{w}_y & (\cosh(\omega) - 1)\hat{w}_x\hat{w}_y & 1 + (\cosh(\omega) - 1)\hat{w}_y^2 & (\cosh(\omega) - 1)\hat{w}_y\hat{w}_z \\ -\sinh(\omega)\hat{w}_z & (\cosh(\omega) - 1)\hat{w}_x\hat{w}_z & (\cosh(\omega) - 1)\hat{w}_y\hat{w}_z & 1 + (\cosh(\omega) - 1)\hat{w}_z^2 \end{bmatrix} \tag{5.8}$$

The free dynamical equations of the four fermion species are generated by $L_B(C)$ boosts of a free Dirac equation for a particle at rest. They fulfill the *requirement that the boosted particle energy is real.*[46] *A boost* can be performed in momentum space and a coordinate space field equation generated from the momentum space version.

5.2.1 Dirac Equation – Free Charged Lepton Dynamical Equation

In this section we describe a well-known method of obtaining the Dirac equation by conventional Lorentz boosts[47] of the spinor wave function of a particle at rest. A generic positive energy plane wave solution of the Dirac equation for a particle at rest with rest energy m is

$$\psi(x) = e^{-imt}w(0) \tag{5.9}$$

---

[45] This limitation will be required to boost particles at rest to particles having real energies as we will see shortly.

[46] If the energy of a free fundamental particle were not real it would imply the particle was not fundamental but a resonace of some sort that could decay into ???. Thus real energies are a requirement.

[47] A Lorentz boost is a special case of an $L_B(C)$ boost with $\omega_i = 0$ and thus a real $\omega = \omega_r$ and a real $\hat{\mathbf{w}}$.

with w(0) a four component spinor column vector. It satisfies the momentum space Dirac equation for a particle at rest:

$$(m\gamma^0 - m)e^{-imt}w(0) = 0 \tag{5.10}$$

The 4 x 4 spinor matrix form of a Lorentz transformation with relative velocity **v** of the Dirac matrices is

$$S^{-1}(\Lambda(\mathbf{v}))\gamma^\nu S(\Lambda(\mathbf{v})) = \Lambda^\nu{}_\mu(\mathbf{v})\gamma^\mu \tag{5.11}$$

where S(Λ(**v**)) is

$$S(\Lambda(\mathbf{v})) = \exp(-i\omega\sigma_{0i}v_i/(2|\mathbf{v}|)) = \exp(-\omega\gamma^0\boldsymbol{\gamma}\cdot\mathbf{v}/(2|\mathbf{v}|))$$

$$= \cosh(\omega/2)I + \sinh(\omega/2)\gamma^0\boldsymbol{\gamma}\cdot\mathbf{p}/|\mathbf{p}| \tag{5.12}$$

with real $\omega = \text{arctanh}(|\mathbf{v}|)$ and real **v** where $\cosh(\omega/2) = [(E+m)/(2m)]^{½}$ and $\sinh(\omega/2) = |\mathbf{p}|[2m(E+m)]^{-½}$. Also

$$S^{-1}(\Lambda(\mathbf{v})) = \gamma^0 S^\dagger(\Lambda(\mathbf{v}))\gamma^0 = \exp(\omega\gamma^0\boldsymbol{\gamma}\cdot\mathbf{v}/(2|\mathbf{v}|))$$

$$= \cosh(\omega/2)I - \sinh(\omega/2)\gamma^0\boldsymbol{\gamma}\cdot\mathbf{p}/|\mathbf{p}| \tag{5.13}$$

If we now apply S(Λ(**v**)) to the momentum space Dirac equation of a particle at rest (eq. 5.10) we find

$$0 = S(\Lambda(\mathbf{v}))(m\gamma^0 - m)e^{-imt}w(0)$$
$$= [mS(\Lambda(\mathbf{v}))\gamma^0 S^{-1}(\Lambda(\mathbf{v})) - m]S(\Lambda(\mathbf{v}))w(0)$$

A straightforward evaluation shows

$$mS(\Lambda(v))\gamma^0 S^{-1}(\Lambda(v)) = g_{\mu\nu}p^\mu\gamma^\nu = \not{p} \tag{5.14}$$

where $p^0 = (p^2 + m^2)^{½}$, $\mathbf{p} = \gamma m\mathbf{v}$, and $p = |\mathbf{p}|$. In addition we define

$$S(\Lambda(v))w(0) = w(p) \tag{5.15}$$

a positive energy Dirac spinor. Therefore the Dirac equation in momentum space has the familiar form:

$$(\not{p} - m)e^{-ip \cdot x}w(p) = 0 \qquad (5.16)$$

where the exponential factor, mt, is also boosted to $p \cdot x$. Eq. 5.16 implies the free, coordinate space Dirac equation:

$$(i\gamma^{\mu}\partial/\partial x^{\mu} - m)\psi(x) = 0 \qquad (5.17)$$

We identify this equation as the dynamical equation of a free lepton.

5.2.2 Form of a $L_B(C)$ Dirac γ matrix Boost

The form of the $L_B(C)$ Dirac matrix boost transformation corresponding to the coordinate transformation eq. 5.8 is:

$$\begin{aligned} S_C(\omega, \mathbf{v_c}) \equiv S_C &= \exp(-i\omega\sigma_{0k}\hat{w}_k/2) = \exp(-\omega\gamma^0\boldsymbol{\gamma}\cdot\hat{\mathbf{w}}/2) \\ &= \cosh(\omega/2)I + \sinh(\omega/2)\gamma^0\boldsymbol{\gamma}\cdot\hat{\mathbf{w}} \end{aligned} \qquad (5.18)$$

with $\mathbf{v_c}$ and $\hat{\mathbf{w}}$ defined by eqns. 5.7 and 5.5 respectively. The inverse transformation is

$$\begin{aligned} S_C^{-1}(\omega, \mathbf{v_c}) &= \gamma^2\gamma^0 K^{-1} S_C^{\dagger} K \gamma^0\gamma^2 = \gamma^2\gamma^0 S_C^{\ T} \gamma^0\gamma^2 = \exp(\omega\gamma^0\boldsymbol{\gamma}\cdot\hat{\mathbf{w}}/2) \\ &= \cosh(\omega/2)I - \sinh(\omega/2)\gamma^0\boldsymbol{\gamma}\cdot\hat{\mathbf{w}} \end{aligned} \qquad (5.19)$$

where the superscript T denotes the transpose and *K* is the complex conjugation operator (that also appears in the time-reversal operator). Note that S_C is not unitary just as $S(\Lambda(\mathbf{v}))$ of eq. 5.12 is not unitary.

We now apply an $L_B(C)$ spinor boost (eq. 5.18) to the Dirac equation for a particle at rest in this more general case of complex ω and $\hat{\mathbf{w}}$.

$$\begin{aligned} 0 &= S_C(\omega, \mathbf{v_c}))(m\gamma^0 - m)e^{-imt}w(0) \\ &= [mS_C\gamma^0 S_C^{-1} - m]e^{-imt}S_C w(0) \end{aligned} \qquad (5.20)$$

where $S_C = S_C(\omega, \mathbf{v}_c)$. After some algebra we find

$$mS_C\gamma^0 S_C^{-1} = m[\cosh(\omega)\gamma^0 - \sinh(\omega)\boldsymbol{\gamma}\cdot\hat{\mathbf{w}}] \tag{5.21}$$

We will use these complex boosts to generate the other fermion species' Dirac-like equations.

5.2.3 Tachyon Dirac Equation – Free Neutral Leptons

The development of the $L_B(C)$ spinor boost transformation (eqns. 5.18 – 5.21) leads to two possible forms of the tachyon Dirac equation. One form will lead to a free lagrangian theory with physical left-handed neutrinos. The other form leads to a free lagrangian theory with physical right-handed neutrinos.

5.2.3.1 Form Leading to a Left-Handed Neutrino

If the real and imaginary relative vectors parts of $\hat{\mathbf{w}}$, namely $\hat{\mathbf{u}}_r$ and $\hat{\mathbf{u}}_i$, are parallel, then $\hat{\mathbf{u}}_r\cdot\hat{\mathbf{u}}_i = 1$ and

$$\omega = \omega_r + i\omega_i \tag{5.22}$$

Eqns. 5.18 and 5.19 then imply

$$mS_C\gamma^0 S_C^{-1} = m[\cosh(\omega_r)\cos(\omega_i) + i\sinh(\omega_r)\sin(\omega_i)]\gamma^0 -$$

$$- m[\sinh(\omega_r)\cos(\omega_i) + i\cosh(\omega_r)\sin(\omega_i)]\boldsymbol{\gamma}\cdot\hat{\mathbf{u}}_r \tag{5.23}$$

or

$$mS_C\gamma^0 S_C^{-1} = \cos(\omega_i)\gamma\cdot p_r + i\sin(\omega_i)\gamma\cdot p_i \tag{5.24}$$

where

$$p_r^{\mathbf{0}} = m\cosh(\omega_r) \qquad p_i^{\mathbf{0}} = m\sinh(\omega_r) \tag{5.25}$$

and

$$\mathbf{p}_r = m\hat{\mathbf{u}}_r\sinh(\omega_r) \qquad \mathbf{p}_i = m\hat{\mathbf{u}}_r\cosh(\omega_r) \tag{5.26}$$

If $\omega_i = 0$, then we recover the momentum space Dirac equation eq. 5.16. If $\omega_i = \pi/2$, then we get the left-handed momentum space tachyon equation:

$$mS_C\gamma^0 S_C^{-1} = i\gamma\cdot p_i \tag{5.27}$$

and the tachyon energy and momentum expressions

$$\mathbf{p} = m\mathbf{v}\gamma_s \qquad E = m\gamma_s \tag{5.28}$$

where $\sinh(\omega) = \gamma_s = (\beta^2 - 1)^{-1/2}$ with $\beta = v/c > 1$. Also

$$S_C w(0) = w_C(p) \tag{5.29}$$

is a tachyon spinor. (See Blaha (2007b).)

The momentum space tachyonic Dirac equation is

$$(i\not{p} - m)e^{ip\cdot x}w_T(p) = 0 \tag{5.30}$$

where $p\cdot x = Et - \mathbf{p\cdot x}$ after performing a corresponding L(C) boost in the exponential factor. Thus the positive energy wave is transformed into a negative energy wave by the superluminal boost transformation.

If we apply $i\not{p}$ to we find the tachyon mass condition is satisfied

$$-E^2 + \mathbf{p}^2 = m^2 \tag{5.31}$$

Transforming back to coordinate space we obtain the "left-handed" *tachyonic Dirac equation*:

$$(\gamma^\mu\partial/\partial x^\mu - m)\psi_T(x) = 0 \tag{5.32}$$

5.2.3.2: Form Leading to a Right-handed Neutrino

If the real and imaginary relative vectors parts of $\hat{\mathbf{w}}$, $\hat{\mathbf{u}}_r$ and $\hat{\mathbf{u}}_i$, are anti-parallel $\hat{\mathbf{u}}_r = -\hat{\mathbf{u}}_i$, then $\hat{\mathbf{u}}_r\cdot\hat{\mathbf{u}}_i = -1$ and

$$\omega = \omega_r - i\omega_i \tag{5.33}$$

Then

$$mS_C\gamma^0 S_C^{-1} = m[\cosh(\omega_r)\cos(\omega_i) - i\sinh(\omega_r)\sin(\omega_i)]\gamma^0 -$$

$$- m[\sinh(\omega_r)\cos(\omega_i) - i\cosh(\omega_r)\sin(\omega_i)]\boldsymbol{\gamma}\cdot\hat{\mathbf{u}}_r \tag{5.34}$$

or

$$mS_C\gamma^0 S_C^{-1} = \cos(\omega_i)\gamma\cdot\mathbf{p}_r - i\sin(\omega_i)\gamma\cdot\mathbf{p}_i \tag{5.35}$$

where

$$p_r^0 = m\cosh(\omega_r) \qquad p_i^0 = m\sinh(\omega_r) \tag{5.36}$$

and

$$\mathbf{p}_r = m\hat{\mathbf{u}}_r\sinh(\omega_r) \qquad \mathbf{p}_i = m\hat{\mathbf{u}}_r\cosh(\omega_r) \tag{5.37}$$

If $\omega_i = \pi/2$, then we obtain the right-handed momentum space tachyon equation.[48]

$$(-\gamma^\mu\partial/\partial x^\mu - m)\psi_T(x) = 0 \tag{5.38}$$

The sign difference between eqns. 5.32 and 5.38 is significant taking account of the required sign of the mass term in the lagrangian due to positivity requirements of the associated hamiltonian. It leads to the difference between a theory with physical interacting, left-handed neutrinos and a theory with physical interacting, right-handed neutrinos as we shall see.

5.2.4 Quark Dirac Equation - Free Up-type Quark Dynamical Equation

There are two other $L_B(C)$ boost cases where we can obtain fermion dynamical equations with a *real* time variable and real energy. In one case $\hat{\mathbf{u}}_r\cdot\hat{\mathbf{u}}_i = 0$ and ω is real.

If the real and imaginary vector parts of $\hat{\mathbf{w}}$, namely $\hat{\mathbf{u}}_r$ and $\hat{\mathbf{u}}_i$, are perpendicular, $\hat{\mathbf{u}}_r\cdot\hat{\mathbf{u}}_i = 0$, then by eq. 5.4

$$\omega = (\omega_r^2 - \omega_i^2)^{1/2} \tag{5.39}$$

Thus ω is either pure real ($\omega_r \geq \omega_i$) or pure imaginary ($\omega_r < \omega_i$).

The momentum space equation generated by the corresponding $L_B(C)$ spinor boost (eqns. 5.12 and 5.13) is

---

[48] We note that $\gamma_s = (\beta^2 - 1)^{-1/2}$, if expressed in terms of ω, has a branch cut extending from $<-\infty, +\infty>$ in the complex ω plane. Thus values of ω with positive imaginary parts are physically different from values of ω with negative imaginary parts (different sheets).

$$\{m \cosh(\omega)\gamma^0 - m \sinh(\omega)\boldsymbol{\gamma}\cdot(\omega_r\hat{\mathbf{u}}_r + i\omega_i\hat{\mathbf{u}}_i)/\omega - m\}e^{-ip\cdot x}w_c(p) = 0 \qquad (5.40)$$

Defining the momentum 4-vector

$$p = (p^0, \mathbf{p}) \qquad (5.41)$$

where

$$p^0 = m \cosh(\omega) \qquad \mathbf{p} = \mathbf{p}_r + i\mathbf{p}_i \qquad (5.42)$$

with

$$\mathbf{p}_r = m\omega_r\hat{\mathbf{u}}_r \sinh(\omega)/\omega \qquad \mathbf{p}_i = m\omega_i\hat{\mathbf{u}}_i \sinh(\omega)/\omega \qquad (5.43)$$

$$\mathbf{p}_r\cdot\mathbf{p}_i = 0 \qquad (5.44)$$

then we obtain a positive energy Dirac-like equation

$$[p\cdot\gamma - m]e^{-ip\cdot x}w_c(p) = 0$$

or

$$[p^0\gamma^0 - (\mathbf{p}_r + i\mathbf{p}_i)\cdot\boldsymbol{\gamma} - m]e^{-ip\cdot x}w_c(p) = 0 \qquad (5.45)$$

with a complex 3-momentum $\mathbf{p}$ and the 4-momentum mass shell condition:

$$p^2 = p^{0\,2} - \mathbf{p}_r\cdot\mathbf{p}_r + \mathbf{p}_i\cdot\mathbf{p}_i = m^2 \qquad (5.46)$$

Note

$$|\mathbf{v}_c| = |\mathbf{p}|/p^0 = [(\mathbf{p}_r + i\mathbf{p}_i)\cdot(\mathbf{p}_r + i\mathbf{p}_i)]^{1/2}/p^0 = \tanh(\omega) \qquad (5.47)$$

and so the Lorentz factor is

$$\gamma = \cosh(\omega) \qquad (5.48)$$

Eq. 5.45 is the momentum space equivalent of the wave equation

$$[i\gamma^0\partial/\partial t + i\boldsymbol{\gamma}\cdot(\nabla_r + i\nabla_i) - m]\psi_{Cu}(t, \mathbf{x}_r, \mathbf{x}_i) = 0 \qquad (5.49)$$

where $\mathbf{x} = \mathbf{x}_r - i\mathbf{x}_i$, and where the grad operators ∇_r and ∇_i are with respect to $\mathbf{x}_r$ and $\mathbf{x}_i$ respectively. Since $\hat{\mathbf{u}}_r\cdot\hat{\mathbf{u}}_i = 0$, which in turn implies eq. 5.44, we see that there is a subsidiary condition on the wave function

$$\nabla_r \cdot \nabla_i \, \psi_{Cu}(t, \mathbf{x}_r, \mathbf{x}_i) = 0 \qquad (5.50)$$

We will call the particles satisfying eqns. 5.49 and 5.50 *complexons.*

We note that eq. 5.49 is covariant under the real Lorentz group. Eq. 5.50 can be easily put into a covariant form as the difference of two 4-vectors squared (which is a Lorentz group invariant):

$$[\gamma^0\partial/\partial t + i\boldsymbol{\gamma}\cdot(\nabla_r + i\nabla_i)]^2 - [\gamma^0\partial/\partial t + i\boldsymbol{\gamma}\cdot(\nabla_r - i\nabla_i)]^2 = 4\nabla_r\cdot\nabla_i$$

We identify eq. 5.49 as the dynamical equation of a free, up-type quark.[49] Later we will show that the fourier solution of the equation has an SU(3) global symmetry.

5.2.5 Left-handed Quark Dirac Equation – Free Down-type Quark Dynamical Equation

In this case we set $\hat{\mathbf{u}}_r\cdot\hat{\mathbf{u}}_i = 0$. Then by eq. 5.4

$$\omega = (\omega_r^2 - \omega_i^2)^{½}$$

Thus ω again starts out either purely real (if $\omega_r \geq \omega_i$) or purely imaginary (if $\omega_r < \omega_i$). In this case we also choose ω real, and then change ω to

$$\omega = (\omega_r^2 - \omega_i^2)^{½} \rightarrow \omega' = (\omega_r^2 - \omega_i^2)^{½} + i\pi/2 = \omega + i\pi/2$$

by adding iπ/2 to ω in eq. 15.65 since ω is a free parameter and proceed as we did in the prior tachyon case.[50] The resulting $L_B(C)$ boost

$$\Lambda_C = \exp[i((\omega_r^2 - \omega_i^2)^{½} + i\pi/2)(\omega_r\hat{\mathbf{u}}_r + i\omega_i\hat{\mathbf{u}}_i)\cdot\mathbf{K}/\omega] \qquad (5.51)$$

---

[49] As we point out later in the chapter on experimental questions we have chosen to consider the possibility that the free dynamical equations for up-type and down-type quarks are the same as those for charged leptons and neutral leptons respectively, OR the possibility that quarks are complexons. The global SU(3) symmetry implicit in complexons strongly suggested that quarks be identified with complexons.

[50] Here again the choice of ω in eq. 5.51 leads to a left-handed quark ElectroWeak sector while the choice ω' = ω – iπ/2 leads to a right-handed quark ElectroWeak sector.

becomes a left-handed quark-like boost. The consequent tachyon dynamical equation is

$$[\gamma^0\partial/\partial t + \boldsymbol{\gamma}\cdot(\nabla_r + i\nabla_i) - m]\psi_{Cd}(x) = 0 \qquad (5.52)$$

with the constraint equation

$$\nabla_r\cdot\nabla_i\ \psi_{Cd}(t, \mathbf{x}_r, \mathbf{x}_i) = 0 \qquad (5.53)$$

We will call the particles satisfying eqns. 5.52 and 5.53 *tachyonic complexons*.

We identify eq. 5.52 as the dynamical equation of a free, down-type, left-handed quark. Later we will show that the fourier solution of this equation also has an SU(3) global symmetry.

Thus we have seen how to generate the dynamical equations of the four fermion species. It is remarkable that the set of complex Lorentz group boosts, limited to those that boost a particle from rest to one with a real-valued energy, generates exactly four types of fermions which nicely match a major feature of the fermion particle spectrum.

5.3 Free Lagrangians of the Four Fermion Species

In defining the lagrangians for the four species that lead to their dynamical equations in the usual canonical way, we require the conventional quantum field theory feature that the hamiltonian derived from the lagrangian is hermitean. We will develop a separate lagrangian for each species and combine them later in the derivation of the form of the Standard Model. The interested readers can derive these lagrangians for themselves or see their derivation in Blaha (2007b) or part 4 of Blaha (2011c).

5.3.1 Free Charged Lepton Lagrangian

The Dirac equation lagrangian is

$$\mathcal{L} = \bar{\psi}(i\gamma^\mu\partial/\partial x^\mu - m)\psi(x) \qquad (5.54)$$

where

$$\bar{\psi} = \psi^\dagger\gamma^0$$

and $\psi^\dagger$ is the hermitean conjugate of ψ.

5.3.2 Free Neutral Lepton Lagrangian

$$\mathscr{L}_T = \psi_T{}^S(\gamma^\mu\partial/\partial x^\mu - m)\psi_T(x) \tag{5.55}$$

where

$$\psi_T{}^S = \psi_T{}^\dagger\, i\gamma^0\gamma^5 \tag{5.56}$$

The peculiar form of the tachyon lagrangian is necessitated by the required hermiticity of the hamiltonian calculated from it.

5.3.3 Free Up-type Quark Lagrangian

The lagrangian density of a free up-type complexon quark is

$$\mathscr{L}_{Cu} = \bar{\psi}_{Cu}(i\gamma^\mu D_\mu - m)\psi_{Cu}(x) \tag{5.57}$$

where $\bar{\psi}_{Cu} = \psi_{Cu}{}^\dagger\gamma^0$ and

$$\psi_{Cu}{}^\dagger = [\psi_{Cu}(\mathbf{x}_r, \mathbf{x}_i)]^\dagger\, |_{\mathbf{x}_i = -\mathbf{x}_i} \tag{5.58}$$

$$D_0 = \partial/\partial x^0$$
$$D_k = \partial/\partial x_r{}^k + i\,\partial/\partial x_i{}^k \tag{5.59}$$

for k = 1, 2, 3. The action

$$I = \int d^7x \mathscr{L}_{Cu} \tag{5.60}$$

is invariant under real Lorentz transformations involving $(x^0, \mathbf{x}_r)$ with $\mathbf{x}_i$ held constant. It is easy to show that this action is also real.

5.3.4 Free Down-type Quark Lagrangian

The simplest, physically acceptable, free, left-handed tachyon quark lagrangian density is:

$$\mathscr{L}_{Cd} = \psi_{Cd}{}^C(x)(\gamma^0\partial/\partial t + \boldsymbol{\gamma}\cdot(\nabla_r + i\nabla_i) - m)\psi_{Cd}(x) \tag{5.61}$$

where

$$\psi_{Cd}{}^{C}(x) = [\psi_{Cd}(x)]^{\dagger}|_{\mathbf{x}_i = -\mathbf{x}_i}\, i\gamma^0\gamma^5 \qquad (5.62)$$

In words, eq. 5.62 states: take the hermitean conjugate of $\psi_{Cd}(x)$; change $\mathbf{x}_i$ to $-\mathbf{x}_i$; and then post-multiply by the indicated factors.

The free tachyon complexon, action

$$I = \int d^7x \mathcal{L}_{Cd} \qquad (5.63)$$

is invariant under real Lorentz transformations involving (x^0, $\mathbf{x}_r$) with $\mathbf{x}_i$ held constant. And the action can be shown to be real.

5.4 Canonical Commutation Relations of the Four Fermion Species

5.4.1 Free Charged Lepton Commutation Relations

The canonically conjugate momentum determined by the lagrangian density eq. 5.54 is

$$\pi_a = \partial\mathcal{L}/\partial\dot{\psi}_a \equiv \partial\mathcal{L}/\partial(\partial\psi_a/\partial t) = i\psi^{\dagger}{}_a \qquad (5.64)$$

The resulting non-zero, canonical, equal time, anti-commutation relations are

$$\{\pi_a(x), \psi_b(x')\} = i\,\delta_{ab}\,\delta^3(x - x')$$

or

$$\{\psi^{\dagger}{}_a(x), \psi_b(x')\} = \delta_{ab}\,\delta^3(x - x') \qquad (5.65)$$

5.4.2 Free Neutral Lepton Commutation Relations

We identify neutral leptons with tachyons. Having defined a suitable tachyon lagrangian we can now perform its canonical quantization. The conjugate momentum calculated from the lagrangian density eq. 5.55 is

$$\pi_{Ta} = \partial\mathcal{L}_T/\partial\dot{\psi}_{Ta} \equiv \partial\mathcal{L}_T/\partial(\partial\psi_{Ta}/\partial t) = -i(\psi_T{}^{\dagger}\gamma^5)_a \qquad (5.66)$$

The resulting non-zero, canonical anti-commutation relations are

$$\{\pi_{T_a}(x),\, \psi_{Tb}(x')\} = i\,\delta_{ab}\,\delta^3(x - x')$$

or

$$\{\psi^\dagger_{T_a}(x),\, \psi_{Tb}(x')\} = -\,[\gamma^5]_{ab}\,\delta^3(x - x') \qquad (5.67)$$

The presence of the γ^5 in eq. 5.67 indicates that parity differentiates between the left-handed and right-handed field anti-commutation relations. We define left-handed and right-handed fields using a transformed set of Dirac matrices:

$$\gamma^0 = \begin{bmatrix} 0 & -I \\ -I & 0 \end{bmatrix} \quad \gamma^i = \begin{bmatrix} 0 & \sigma_i \\ -\sigma_i & 0 \end{bmatrix} \quad \gamma^5 = \begin{bmatrix} I & 0 \\ 0 & -I \end{bmatrix} \qquad (5.68)$$

which are obtained from the usual Dirac matrices by applying the unitary transformation $U = 2^{-½}(I + \gamma^5\gamma^0)$. Note I is the 4×4 identity matrix. The γ^5 chirality operator's eigenvalues define handedness: +1 corresponds to right-handed, and –1 corresponds to left-handed:

$$\gamma^5\psi_{TL} = -\psi_{TL} \qquad \gamma^5\psi_{TR} = \psi_{TR} \qquad (5.69a)$$

The projection operators:

$$\begin{aligned} C^\pm &= ½(I \pm \gamma^5) \\ C^+ + C^- &= I \\ C^{\pm\,2} &= C^\pm \\ C^+C^- &= 0 \end{aligned} \qquad (5.69b)$$

are used to define left and right handed tachyon fields

$$\begin{aligned} \psi_{TL} &= C^-\psi_T \\ \psi_{TR} &= C^+\psi_T \end{aligned} \qquad (5.70)$$

Eq. 5.67 implies the anti-commutation relations

$$\{\psi_{TLa}{}^{\dagger}(x),\ \psi_{TLb}(x')\} = C^{-}{}_{ab}\delta^{3}(x - x') \tag{5.71}$$

$$\{\psi_{TRa}{}^{\dagger}(x),\ \psi_{TRb}(x')\} = -\,C^{+}{}_{ab}\delta^{3}(x - x') \tag{5.72}$$

$$\{\psi_{TLa}{}^{\dagger}(x),\ \psi_{TRb}(x')\} = \{\psi_{TRa}{}^{\dagger}(x),\ \psi_{TLb}(x')\} = 0 \tag{5.73}$$

The lagrangian density of eq. 5.55 decomposes into left-handed and right-handed parts (modulo the mass term):

$$\mathscr{L}_{T} = \psi_{TL}{}^{\dagger}\gamma^{0}i\gamma^{\mu}\partial_{\mu}\psi_{TL} - \psi_{TR}{}^{\dagger}\gamma^{0}i\gamma^{\mu}\partial_{\mu}\psi_{TR} - im[\psi_{TR}{}^{\dagger}\gamma^{0}\psi_{TL} - \psi_{TL}{}^{\dagger}\gamma^{0}\psi_{TR}] \tag{5.74}$$

Noting the sign difference between the left-handed and right-handed kinetic terms in eq. 5.74, we see that the right-handed kinetic term (having the wrong sign) leads to an unphysical quantum field theory. This difference is the source of parity violation in the Standard Model, and the source of its left-handedness, as we will show in a later section.

5.4.3 Free Up-type Quark Commutation Relations

There are two choices for the up-type free quark (and correspondingly two choices in the interacting quark case). The first choice is to assume a free up-type quark theory similar to the theory of the free charged lepton considered earlier in this section based on the lagrangian eq. 5.54. Pursuing this approach would lead to the conventional Standard Model quark up-type sector.[51]

However it is clear there is a profound difference between quarks and leptons in nature. Quarks are always bound; leptons are not necessarily bound. Quarks experience the strong interaction; leptons do not. Consequently it is of interest to consider a second case embodied in the third type of fermion as specified in eqns. 5.57–5.60. In this case we shall see that the field solutions contain an embedded global SU(3) symmetry. This symmetry leads us to postulate that quarks are in the <u>3</u> representation of an SU(3) symmetry that we identify with color for normal particles and the <u>1</u> singlet representation for Dark quarks.

---

[51] This possibility results in what we call The Standard Standard Model.

In this second case the conjugate momentum is

$$\pi_{Cua} = \partial\mathcal{L}/\partial\dot{\psi}_{Cua} \equiv \partial\mathcal{L}/\partial(\partial\psi_{Cua}/\partial x^0) = i\psi_{Cu\,a}^{\dagger} \tag{5.75}$$

where a is a spinor index. The non-zero anti-commutation relation, *at first glance*, appears to be

$$\{\psi_{Cu\,\alpha}^{\dagger}(x), \psi_{Cub}(y)\} = \delta_{ab}\,\delta^3(x_r - y_r)\delta^3(x_i - y_i) \tag{5.76}$$

However the constraint eq. 5.50 *requires* the anti-commutator to be

$$\{\psi_{Cu\,\alpha}^{\dagger}(x), \psi_{Cub}(y)\} = -\,\delta_{ab}\delta'(\nabla_r\cdot\nabla_i/m^2)[\delta^3(x_r - y_r)\delta^3(x_i - y_i)] \tag{5.77}$$

where all ∇_r and ∇_i are ∇ derivatives with respect to x, and where $\delta'(\nabla_r\cdot\nabla_i)$ is the derivative of a delta function with the argument being differential operators such as those in eq. 5.50. The minus sign is due to the presence of a *derivative* of a delta-function and is not an issue.

5.4.4 Free Down-type Quark Commutation Relations

The conjugate momentum can be calculated from the lagrangian density eq. 5.61:

$$\pi_{Cda} = \partial\mathcal{L}_{Cd}/\partial\dot{\psi}_{Cda} \equiv \partial\mathcal{L}_{Cd}/\partial(\partial\psi_{Cda}/\partial t) = -i([\psi_{Cd}(x)]^{\dagger}|_{\mathbf{X}_i = -\mathbf{X}_i}\gamma^5)_a \tag{5.78}$$

The resulting non-zero, canonical anti-commutation relations are presumably

$$\{\pi_{Cda}(x), \psi_{Cdb}(y)\} = i\,\delta_{ab}\,\delta^3(x_r - y_r)\delta^3(x_i - y_i) \tag{5.79}$$

based on locality in both the real and imaginary coordinates. However, on taking account of the constraint eq. 5.53 it must be modified to

$$\{\psi_{Cd\,\alpha}^{\dagger}(x)|_{\mathbf{X}_i = -\mathbf{X}_i}, \psi_{Cdb}(y)\} = [\gamma^5]_{ab}\delta'(\nabla_r\cdot\nabla_i/m^2)[\delta^6(x - y)] \tag{5.80}$$

As in the case of the neutral lepton commutation relations the presence of a γ^5 in eq. 5.80 differentiates between left and right-handed field anti-commutators:

$$\{\psi_{CdL\alpha}{}^{\dagger}(x)|_{\mathbf{X_i} = -\mathbf{X_i}'}, \psi_{CdLb}(y)\} = -C^{-}{}_{ab}\delta'(\nabla_r\cdot\nabla_i/m^2)[\delta^6(x-y)] \quad (5.81)$$

$$\{\psi_{CdR\alpha}{}^{\dagger}(x)|_{\mathbf{X_i} = -\mathbf{X_i}'}, \psi_{CdRb}(y)\} = C^{+}{}_{ab}\delta'(\nabla_r\cdot\nabla_i/m^2)[\delta^6(x-y)] \quad (5.82)$$

$$\{\psi_{CdL\alpha}{}^{\dagger}(x)|_{\mathbf{X_i} = -\mathbf{X_i}'}, \psi_{CdRb}(y)\} = \{\psi_{CdR\alpha}{}^{\dagger}(x)|_{\mathbf{X_i} = -\mathbf{X_i}'}, \psi_{CdLb}(x')\} = 0 \quad (5.83)$$

The left-handed fields again have a physically acceptable sign and lead to the left-handedness of the Standard Model quark sector.

5.5 Fourier Expansions of Free Fermion Fields

In this section we will list the fourier expansions, and some salient features, of the four species of fermion fields. The interested reader can read about the details of the fourier expansions in Blaha (2007b). We will express the fourier expansions in terms of light front variables[52] for the Dirac and tachyon species although they are required only for a correct canonical *tachyon* quantization.[53] (If equal-time canonical quantization of tachyons is attempted, then non-localities,[54] and other problems, appear.) The light front variables are:

$$x^{\pm} = (x^0 \pm x^3)/\surd 2 \qquad (5.84)$$
$$\partial/\partial x^{\pm} \equiv \partial^{\mp} \equiv (\partial/\partial x^0 \pm \partial/\partial x^3)/\surd 2$$

with the "transverse" coordinate variables, x^1 and x^2, unchanged.

---

[52] L. Susskind, Phys. Rev. **165**, 1535 (1968); K. Bardakci and M. B. Halpern Phys. Rev. **176**, 1686 (1968), S. Weinberg, Phys. Rev. **150**, 1313 (1966); J. Kogut and D. Soper, Phys. Rev. **D1**, 2901 (1970); J. D. Bjorken, J. Kogut, and D. Soper, Phys. Rev. **D3**, 1382 (1971); R. A. Neville and F. Rohrlich, Nuov. Cim. **A1**, 625 (1971); F. Rohrlich, Acta Phys Austr. Suppl. **8**, 277 (1971); S-J Chang, R. Root, and T-M Yan, Phys. Rev. **D7**, 1133 (1973); S-J Chang, and T-M Yan, Phys. Rev. **D7**, 1147 (1973); T-M Yan, Phys. Rev. **D7**, 1761 (1973); T-M Yan, Phys. Rev. **D7**, 1780 (1973); C. Thorn, Phys. Rev. **D19**, 639 (1979); and references therein.

[53] This was first demonstrated in Blaha (2006) who showed tachyon fields must be separated into left-handed and right-handed parts, and then second quantized using light-front coordinates, to obtain local, equal light-front anti-commutators.

[54] See G. Feinberg, Phys. Rev. **159**, 1089 (1967) for example.

5.5.1 Free Charged Lepton Field Fourier Expansion

The fourier expansion of the free charged lepton field is the conventional Dirac field expansion. In light-front coordinates the free, "+" wave function Fourier expansion[55] of a Dirac field is:

$$\psi^+(x) = \sum_{\pm s} \int d^2p dp^+ N^+(p)\theta(p^+)[b^+(p, s)u^+(p, s)e^{-ip\cdot x} + d^{+\dagger}(p, s)v^+(p, s)e^{+ip\cdot x}] \tag{5.85}$$

where

$$N^+(p) = [m/((2\pi)^3 p^+)]^{1/2} \tag{5.86}$$

with $u^+(p, s)$ and $v^+(p, s)$ being projections of the conventional spinors u and v; and with the non-zero creation and annihilation operator anti-commutators:

$$\begin{aligned} \{b^+(q,s), b^{+\dagger}(p,s')\} &= \delta_{ss'}\delta^2(\mathbf{q} - \mathbf{p})\delta(q^+ - p^+) \\ \{d^+(q,s), d^{+\dagger}(p,s')\} &= \delta_{ss'}\delta^2(\mathbf{q} - \mathbf{p})\delta(q^+ - p^+) \end{aligned} \tag{5.87}$$

5.5.2 Free Neutral Lepton Field Fourier Expansion

The free neutral lepton field fourier expansion is somewhat more involved since it is a tachyon field. The free, "+" light-front, *left-handed* tachyon wave function Fourier expansion is:

$$\psi_{TL}^+(x) = \sum_{\pm s} \int d^2p dp^+ N_{TL}^+(p)\theta(p^+)[b_{TL}^+(p, s)u_{TL}^+(p, s)e^{-ip\cdot x} + d_{TL}^{+\dagger}(p, s)v_{TL}^+(p, s)e^{+ip\cdot x}] \tag{5.88}$$

with

$$N_{TL}^+(p) = [2m|\mathbf{p}|/((2\pi)^3(p^+(p^+ - p^-) + p_\perp^{\,2}))]^{1/2} \tag{5.89}$$

and where the non-zero anti-commutators of the Fourier coefficient operators are

$$\begin{aligned} \{b_{TL}^+(q,s), b_{TL}^{+\dagger}(p,s')\} &= \delta_{ss'}\delta^2(\mathbf{q} - \mathbf{p})\delta(q^+ - p^+) \\ \{d_{TL}^+(q,s), d_{TL}^{+\dagger}(p,s')\} &= \delta_{ss'}\delta^2(\mathbf{q} - \mathbf{p})\delta(q^+ - p^+) \end{aligned} \tag{5.90}$$

---

[55] See S-J Chang and T- M. Yan Phys. Rev. D7, (1973) for a detailed presentation on light-front (infinite momentum frame) quantization of Dirac fields as well as Blaha (2007b).

The spinors are

$$u_{TL}^{+}(p, s) = C^{-} R^{+}S_C(\omega, \mathbf{v_c})w^{1}(0)$$
$$u_{TL}^{+}(p, -s) = C^{-} R^{+}S_C(\omega, \mathbf{v_c})w^{2}(0)$$
$$v_{TL}^{+}(p, s) = C^{-} R^{+}S_C(\omega, \mathbf{v_c})w^{3}(0)$$
$$v_{TL}^{+}(p, -s) = C^{-} R^{+}S_C(\omega, \mathbf{v_c})w^{4}(0)$$

where the superscript "T" indicates the transpose, $R^{\pm} = ½(I \pm \gamma^0\gamma^3)$ are ± field light-front projection operators[56], $S_C(\omega, \mathbf{v_c})$ is an $L_B(C)$ spinor boost, and the column 4-vector components $[w^i(0)]^k = \delta^{ik}$.

The case of *right-handed* free neutral lepton fields (tachyons) is similar to the left-handed case with only two differences: a minus sign in the creation and annihilation operator anti-commutation relations, and the use of right-handed projection operators. The right-handed tachyon wave function light-front Fourier expansion has the form:

$$\psi_{TR}^{+}(x) = \sum_{\pm s} \int d^2p dp^+ N_{TR}^{+}(p)\theta(p^+)[b_{TR}^{+}(p, s)u_{TR}^{+}(p, s)e^{-ip\cdot x} + d_{TR}^{+\dagger}(p, s)v_{TR}^{+}(p, s)e^{+ip\cdot x}]$$

with the non-zero anti-commutators of the Fourier coefficient operators:

$$\{b_{TR}^{+}(q,s), b_{TR}^{+\dagger}(p,s')\} = -\delta_{ss'}\delta^2(\mathbf{q} - \mathbf{p})\delta(q^+ - p^+)$$
$$\{d_{TR}^{+}(q,s), d_{TR}^{+\dagger}(p,s')\} = -\delta_{ss'}\delta^2(\mathbf{q} - \mathbf{p})\delta(q^+ - p^+)$$

The right-handed spinors are

$$u_{TR}^{+}(p, s) = C^{+}R^{+} S_C(\omega, \mathbf{v_c})w^{1}(0)$$
$$u_{TR}^{+}(p, -s) = C^{+}R^{+} S_C(\omega, \mathbf{v_c})w^{2}(0)$$
$$v_{TR}^{+}(p, s) = C^{+}R^{+} S_C(\omega, \mathbf{v_c})w^{3}(0)$$
$$v_{TR}^{+}(p, -s) = C^{+}R^{+} S_C(\omega, \mathbf{v_c})w^{4}(0)$$

5.5.3 Free Up-Type Complexon Quark Field Fourier Expansion

We believe quarks are complexons although this is an experimental question that remains to be resolved. Otherwise the quark field fourier

---

[56] Blaha (2007b) p. 48.

expansions would be the same as the lepton fourier expansions described in the previous two sub-sections.

We will express the fourier expansion of the up-type complexon quark field in conventional coordinates for the sake of illustration and because they are more familiar to most physicists. The form of the up-type complexon quark field, ignoring generations and color indices temporarily, and taking account of the subsidiary condition is[57,58]

$$\psi_{Cu}(x_r, x_i) = \sum_{\pm s} \int d^3p_r d^3p_i\, N_C(p)\delta(\mathbf{p}_r\cdot\mathbf{p}_i/m^2)[b_{Cu}(p,s)u_{Cu}(p, s)e^{-i(p\cdot x + p^*\cdot x^*)/2} + $$
$$+ d_{Cu}^{\dagger}(p,s)v_{Cu}(p, s)e^{+i(p\cdot x + p^*\cdot x^*)/2}] \qquad (5.91)$$

where $\mathbf{p} = \mathbf{p}_r + i\mathbf{p}_i$, $\mathbf{x} = \mathbf{x}_r - i\mathbf{x}_i$, $p\cdot x = p^0x^0 - \mathbf{p}\cdot\mathbf{x}$, and where we use

$$(p\cdot x + p^*\cdot x^*)/2 = p^0x^0 - \mathbf{p}_r\cdot\mathbf{x}_r - \mathbf{p}_i\cdot\mathbf{x}_i \qquad (5.92)$$

in the exponentials in order to avoid divergences that would appear in the calculation of the equal-time commutator, the Feynman propagator and other quantities of interest after second quantization. Note that

$$(\nabla_r + i\nabla_i)e^{-i(p\cdot x + p^*\cdot x^*)/2} = i(\mathbf{p}_r + i\mathbf{p}_i)e^{-i(p\cdot x + p^*\cdot x^*)/2} \qquad (5.93)$$

and

$$(\nabla_r + i\nabla_i)e^{-ip^*\cdot x^*} = 0 \qquad (5.94)$$

for all p.

Further we note

$$N_C(p) = [2m/((2\pi)^6p^0)]^{1/2} \qquad (5.95)$$

and

$$u_{Cu}(p, s) = S_C(\omega, \mathbf{v}_c)w^1(0)$$

---

[57] Note that when $|\mathbf{p}_i| \geq |\mathbf{p}_r|$ (for imaginary $\omega = (\omega_r^2 - \omega_i^2)^{1/2}$) the 3-momentum becomes imaginary $\mathbf{p}\cdot\mathbf{p} < 0$. However, since we will be identifying confined quarks with this type of particle – much modified by a confining color quark interaction – the issue of an imaginary 3-momentum in the hypothetical free quark case becomes moot. We note the energy gap between positive and negative energy states disappears so E = 0 is possible. Thus real Lorentz transformations can mix positive and negative energy states. The solution is to do all calculations in the light-front frame as we do for tachyons. Then the mixing issue is resolved. In the present case we second quantize on the "time-front" for illustrative purposes.

[58] We scale $\mathbf{p}_r\cdot\mathbf{p}_i$ with m^2 in the delta function for convenience. In the case of a zero mass particle some other scale could be used.

$$u_{Cu}(p, -s) = S_C(\omega, \mathbf{v_c})w^2(0)$$
$$v_{Cu}(p, s) = S_C(\omega, \mathbf{v_c})w^3(0)$$
$$v_{Cu}(p, -s) = S_C(\omega, \mathbf{v_c})w^4(0) \qquad (5.96)$$

The momentum 4-vector $p = (p^0, \mathbf{p})$ is related to the other quantities by

$$p^0 = m\cosh(\omega) \qquad \mathbf{p} = \mathbf{p_r} + i\mathbf{p_i}$$
$$\mathbf{p_r} = m\omega_r\hat{\mathbf{u}}_r \sinh(\omega)/\omega \qquad \mathbf{p_i} = m\omega_i\hat{\mathbf{u}}_i \sinh(\omega)/\omega$$
$$\omega = (\omega_r^{\ 2} - \omega_i^{\ 2})^{1/2}$$
$$\mathbf{p_r}\cdot\mathbf{p_i} = 0$$
$$\hat{\mathbf{w}} = (\omega_r\hat{\mathbf{u}}_r + i\omega_i\hat{\mathbf{u}}_i)/\omega$$
$$\hat{\mathbf{w}}\cdot\hat{\mathbf{w}} = 1$$
$$\mathbf{v_c} = \hat{\mathbf{w}}\tanh(\omega) \qquad (5.97)$$

The non-zero anti-commutators of the Fourier coefficient operators are

$$\{b_{Cu}(p,s), b_{Cu}^{\dagger}(p'^*,s')\} = \delta_{ss'}\delta^3(\mathbf{p_r} - \mathbf{p'_{r'}})\delta^3(\mathbf{p_i} + \mathbf{p'_{i'}})$$
$$\{d_{Cu}(p,s), d_{Cu}^{\dagger}(p'^*,s')\} = \delta_{ss'}\delta^3(\mathbf{p_r} - \mathbf{p'_{r'}})\delta^3(\mathbf{p_i} + \mathbf{p'_{i'}}) \qquad (5.98)$$

5.5.4 Free Down-Type Complexon Quark Field Fourier Expansion

It is necessary that we perform the fourier expansion of the down-type quark field in light-front coordinates in order to obtain a local canonical quantization.

5.5.4.1 Left-Handed Down-Type Complexon Quark Field Fourier Expansion

The independent, left-handed, down-type, complexon quark field Fourier expansion (free, "+" light-front, left-handed, tachyonic complexon), ignoring fermion generations and color indices temporarily, and taking account of the subsidiary condition, is

$$\psi_{CdL}^{\ +}(x_r, x_i) = \sum_{\pm s}\int d^2p_r dp^+ d^3p_i\, N_{CdL}^{\ +}(p)\theta(p^+)\cdot$$
$$\cdot\delta((p_i^{\ 3}(p^+ - p^-)/\surd 2 + \mathbf{p}_{r\perp}\cdot\mathbf{p}_{i\perp})/m^2)\cdot$$
$$\cdot[b_{CdL}^{\ +}(p, s)u_{CdL}^{\ +}(p, s)e^{-i(p\cdot x + p^*\cdot x^*)/2} +$$
$$+ d_{CdL}^{\ +\dagger}(p, s)v_{CdL}^{\ +}(p, s)e^{+i(p\cdot x + p^*\cdot x^*)/2}] \qquad (5.99)$$

where

$$N_{CdL}{}^{+}(p) = (2\pi)^{-3}(2m/p^{+})^{½}$$

and

$$u_{CdL}{}^{+}(p, s) = C^{-} R^{+} S_{C}(\omega',\mathbf{v_c})w^{1}(0)$$
$$u_{CdL}{}^{+}(p, -s) = C^{-} R^{+} S_{C}(\omega',\mathbf{v_c})w^{2}(0)$$
$$v_{CdL}{}^{+}(p, s) = C^{-} R^{+} S_{C}(\omega',\mathbf{v_c})w^{3}(0)$$
$$v_{CdL}{}^{+}(p, -s) = C^{-} R^{+} S_{C}(\omega',\mathbf{v_c})w^{4}(0) \qquad (5.100)$$

where

$$\omega' = \omega + i\pi/2 \qquad (5.101)$$

The momentum 4-vector $p = (p^{0}, \mathbf{p})$ is related to the other quantities by eq. 5.97 with ω replaced by $\omega' = \omega + i\pi/2$ in all the relations of eq. 5.97.

The non-zero anti-commutators of the Fourier coefficient operators are:

$$\{b_{CdL}(p,s), b_{CdL}{}^{\dagger}(p'^{*},s')\} = 2^{-½}\delta_{ss'}\delta(p^{+} - p'^{+})\delta^{2}(\mathbf{p_r} - \mathbf{p'}_{r'})\delta^{3}(\mathbf{p_i} + \mathbf{p'}_{i'})$$
$$\{d_{CdL}(p,s), d_{CdL}{}^{\dagger}(p'^{*},s')\} = 2^{-½}\delta_{ss'}\delta(p^{+} - p'^{+})\delta^{2}(\mathbf{p_r} - \mathbf{p'}_{r'})\delta^{3}(\mathbf{p_i} + \mathbf{p'}_{i'}) \qquad (5.102)$$

5.5.4.2 Right-Handed Down-Type Complexon Quark Field Fourier Expansion

The independent, right-handed down-type complexon quark field Fourier expansion (free, "+" light-front, right-handed, tachyonic complexon), ignoring fermion generations and color indices temporarily, and taking account of the subsidiary condition, is

$$\psi_{CdR}{}^{+}(x_r, x_i) = \sum_{\pm s} \int d^{2}p_{r}dp^{+}d^{3}p_{i}\, N_{CdR}{}^{+}(p)\theta(p^{+})\delta((p_i{}^{3}(p^{+} - p^{-})/\sqrt{2} + \mathbf{p}_{r\perp}\cdot\mathbf{p}_{i\perp})/m^{2})\cdot$$
$$\cdot[b_{CdR}{}^{+}(p, s)u_{CdR}{}^{+}(p, s)e^{-i(p\cdot x + p^{*}\cdot x^{*})/2} +$$
$$+ d_{CdR}{}^{+\dagger}(p, s)v_{CdR}{}^{+}(p, s)e^{+i(p\cdot x + p^{*}\cdot x^{*})/2}] \qquad (5.103)$$

where

$$N_{CdR}{}^{+}(p) = (2\pi)^{-3}(2m/p^{+})^{½}$$

and

$$u_{CdR}{}^{+}(p, s) = C^{+} R^{+} S_{C}(\omega',\mathbf{v_c})w^{1}(0)$$
$$u_{CdR}{}^{+}(p, -s) = C^{+} R^{+} S_{C}(\omega',\mathbf{v_c})w^{2}(0)$$

$$v_{CdR}{}^{+}(p, s) = C^{+} R^{+} S_C(\omega', \mathbf{v}_c) w^3(0)$$
$$v_{CdR}{}^{+}(p, -s) = C^{+} R^{+} S_C(\omega', \mathbf{v}_c) w^4(0) \qquad (5.104)$$

where

$$\omega' = \omega + i\pi/2 \qquad (5.101)$$

The momentum 4-vector p = (p^0, **p**) is related to the other quantities by eq. 5.97 with ω replaced by ω' = ω + iπ/2 in all the relations of eq. 5.97.

The non-zero anti-commutators of the Fourier coefficient operators are:

$$\{b_{CdR}(p,s),\ b_{CdR}{}^{\dagger}(p'^{*},s')\} = -2^{-1/2}\,\delta_{ss'}\delta(p^{+} - p'^{+})\delta^2(\mathbf{p}_r - \mathbf{p}'_{r'})\delta^3(\mathbf{p}_i + \mathbf{p}'_{i'})$$
$$\{d_{CdR}(p,s),\ d_{CdR}{}^{\dagger}(p'^{*},s')\} = -2^{-1/2}\delta_{ss'}\delta(p^{+} - p'^{+})\delta^2(\mathbf{p}_r - \mathbf{p}'_{r'})\delta^3(\mathbf{p}_i + \mathbf{p}'_{i'}) \qquad (5.105)$$

5.6 The Four Fermion Species Quantization Program Works

The preceding sections show that we have a valid free particle second quantization procedure for each of the four fermion species. With this formalism in hand we can proceed to develop the extended Standard Model that we call The Complexon Standard Model which has SU(3)⊗SU(2)⊗U(1)⊗SU(2)⊗U(1) as its (broken) symmetry group.

6. SU(3)⊗SU(2)⊗U(1)⊗SU(2)⊗U(1) Complexon Standard Model

We derived a set of Standard Models in Blaha (2011c). Since then experiment has revealed new features of Dark Matter that lead us to extend the models to include a Dark sector. The extended Complexon Standard Model which we will simply call The Complexon Standard Model embodies the (broken) symmetry group SU(3)⊗SU(2)⊗U(1)⊗SU(2)⊗U(1).

In section 4.8 of this book we described a quantum coordinate $X^{\mu}(z) = z^{\mu} + i\, Y^{\mu}(z)/M_c^{\,2}$ in eq. 4.21. These coordinates are described in detail in Blaha (2005a) as well as in section 19.5 and part 5 of Blaha (2011c), and chapter 5 of (2012b). We use them to develop Two Tier quantum field theory and show that it eliminates infinities in perturbation calculations to any order.

In our Conplexon Standard Model all fields will be functions of $X^{\mu}(z)$ and consequently all perturbation theory calculations will yield finite results including triangle "anomaly" calculations.

Another issue that appears in any theory with non-Abelian gauge fields is the determination of the Faddeev-Popov Mechanism contributions. We have described the Faddeev-Popov mechanism for complexon gauge fields are described in appendix 19-A of Blaha (2011c). (All the gauge fields in the theory are complexon gauge fields.)

The previous version of the Complexon Standard Model lagrangian is described starting on p. 305 of Blaha (2011c). In this chapter we extend it to explicitly to a new SU(3)⊗SU(2)⊗U(1)⊗SU(2)⊗U(1) form with the Dark sector described in chapter 1.

6.1 The New Complexon Standard Model Lagrangian

In chapter 1 we considered a new SU(2)⊗U(1) sector of the Standard Model that we identified have identified with the symmetry group of Dark Matter. We identified this group as corresponding to one of the subgroups of U(4) – the 4-dimensional Reality group together with the known symmetries of

The Standard Model SU(3)⊗SU(2)⊗U(1). Each of the factors of the direct product corresponds to a subgroup of the Reality group U(4).

With the addition of the SU(2)⊗U(1) group we form the (broken) symmetry group of the new Complexon Standard Model with a Dark Matter sector interacting with the normal matter sector in a minimal way.[59]

6.1.1 Lepton Sector

We begin with the definition of a quadruplet of leptons in chapter 1 – a pair of doublets, one normal and one Dark, instead of a single doublet. We define left and right lepton quadruplets with[60]

$$\Psi_{L,R}(X) = \begin{bmatrix} \psi_{DL,R}(X) \\ \psi_{NL,R}(X) \end{bmatrix} \tag{1.8}$$

where $\psi_{NL,R}(X)$ is a "normal" ElectroWeak-like lepton doublet, and where $\psi_{DL,R}(X)$ is a Dark ElectroWeak-like lepton doublet consisting of a Dark electron-like fermion and a Dark neutrino-like fermion.

We define covariant derivative terms which we express in matrix form are

$$D_{L,R}(X) = \begin{bmatrix} \gamma^{\mu} D_{DL,R\mu} & 0 \\ 0 & \gamma^{\mu} D_{NL,R\mu} \end{bmatrix} \tag{1.10}$$

where the normal matter left-handed covariant derivative is

$$D_{NL\mu} = \partial/\partial X^{\mu} - \tfrac{1}{2} i g \boldsymbol{\sigma}\cdot\mathbf{W}_{\mu} + \tfrac{1}{2} i g' B_{\mu} \tag{1.11}$$

and where the Dark matter left-handed covariant derivative is

---

[59] Based on the three working principles: 1) The only connecting interaction is a weak interaction, 2) The form of ElectroWeak theory remains unchanged, and 3) Dark Matter parallels normal matter in its general characteristics: three (possibly four) generations, SU(3) singlets, an SU(2)⊗U(1) symmetry analogous to ElectroWeak symmetry, SU(2)⊗U(1) dark lepton and dark quark doublets. Equations from chapter 1 retain the same numbering as in that chapter.

[60] X is a quantized coordinate defined by eq. 4.21.

$$D_{DL\mu} = \partial/\partial X^{\mu} - ½ig_D\boldsymbol{\sigma}\cdot\mathbf{W}'_{\mu} + ½ig_D'B'_{\mu} + ½ig_D''B_{\mu} \quad (1.12)$$

with **σ** a vector composed of the Pauli matrices. The right-handed covariant derivatives have a simpler form. The normal matter right-handed covariant derivative is

$$D_{NR\mu} = \partial/\partial X^{\mu} + ig'B_{\mu} \quad (1.13)$$

and the Dark matter right-handed covariant derivative is

$$D_{DR\mu} = \partial/\partial X^{\mu} + ½ig_D'B'_{\mu} + ½ig_D''B_{\mu} \quad (1.14)$$

The normal and Dark electroweak fields in eqns. 1.11-1.14 are functions of X. The Faddeev-Popov mechanism operative for these types of fields is described in appendix 19-A of Blaha (2011c).

6.1.2 Quark Sector

In the Dark *quark* sector we define left and right quark quadruplets with[61]

$$\psi_{qL,R}(X_c) = \begin{bmatrix} \psi_{DqL,R}(X_c) \\ \psi_{NqL,R}(X_c) \end{bmatrix} \quad (1.19)$$

where $\psi_{NqL,R}(X_c)$ is a "normal" ElectroWeak-like quark doublet, and where $\psi_{DqL,R}(X_c)$ is a Dark ElectroWeak-like quark doublet consisting of a Dark quark of unit Dark charge and a Dark quark of zero Dark charge.

The covariant derivative terms are contained in $D_q(X_c)$ which we express in matrix form as

$$D_{qL,R}(X_c) = \begin{bmatrix} \gamma^{\mu}D_{qDL,R\mu}(X_c) & 0 \\ 0 & \gamma^{\mu}D_{qNL,R\mu}(X_c) \end{bmatrix} \quad (1.21)$$

[61] X_c is a quantized coordinate defined by eqns. 4.23a.

where the normal quark matter left-handed covariant derivative is

$$D_{qNL\mu} = \partial/\partial X_c^{\mu} - \tfrac{1}{2}ig\boldsymbol{\sigma}\cdot\mathbf{W}_{\mu} - ig'B_{\mu}/6 \qquad (1.22)$$

and where the Dark quark left-handed covariant derivative is

$$D_{qDL\mu} = \partial/\partial X_c^{\mu} - \tfrac{1}{2}ig_D\boldsymbol{\sigma}\cdot\mathbf{W}'_{\mu} + \tfrac{1}{2}ig_D'B'_{\mu} + \tfrac{1}{2}ig_D''B_{\mu} \qquad (1.23)$$

since Dark quarks are SU(3) singlets with unit or zero Dark charge. The right-handed quark covariant derivatives have a simpler form. The normal quark right-handed covariant derivative is

$$D_{qNR\mu} = \partial/\partial X_c^{\mu} + ig'B_{\mu}/3 \qquad (1.24)$$

and the Dark quark right-handed covariant derivative is

$$D_{qDR\mu} = \partial/\partial X_c^{\mu} + \tfrac{1}{2}ig_D'B'_{\mu} + \tfrac{1}{2}ig_D''B_{\mu} \qquad (1.25)$$

The normal and Dark gauge boson fields in eqns. 1.22-1.25 are functions of X_c. The Faddeev-Popov mechanism operative for these types of fields is described in appendix 19-A of Blaha (2011c).[62] Then the *complexon* quark Standard Model ElectroWeak Sector covariant derivatives in quadruplet matrix form are

$$D_{qL,R}(X_c) = \begin{bmatrix} \gamma^{\mu}D_{qDL,R\mu} & 0 \\ 0 & \gamma^{\mu}D_{qNL,R\mu} \end{bmatrix} \qquad (1.31)$$

The remaining parts of the complexon Standard Model are described in chapter 23 of Blaha (2011) and summarized below. The addition of singlet Dark quark Higgs terms is also required.

---

[62] Those who might be concerned about the propagator term <$W_i(X)$, $W_j(X_c)$> and similar propagators where one field is a function of X and the other field is a function of X_c should note that such terms are to very good approximation equal to <$W_i(X)$, $W_j(X)$> for energies much less than M_c (which could be as large as the Planck energy.)

The lagrangian density and action is[63]

$$\mathcal{L}_{CSM} = \Psi_L^{a\dagger}\gamma^0 i\gamma^\mu D_{L\mu}\Psi_L^a - \Psi_R^{a\dagger}\gamma^0 i\gamma^\mu D_{R\mu}\Psi_{3R}^a + $$
$$+ \Psi_{CL}^{a\dagger}\gamma^0 i\gamma^\mu \mathcal{D}_{qL\mu}\Psi_{CL}^a + \Psi_{CR}^{a\dagger}\gamma^0 i\gamma^\mu \mathcal{D}_{qR\mu}\Psi_{CR}^a -$$
$$- \mathcal{L}_{BareMasses} + \mathcal{L}_{Gauge} + \mathcal{L}_{Mass} \qquad (6.1)$$

where a is the generation index. $\mathcal{L}_{BareMasses}$ contains the fermion bare mass terms. Also,

$$\mathcal{L}_{Gauge} = \mathcal{L}_{GaugeEW} + \mathcal{L}_{GaugeC} \qquad (6.2)$$

with

$$\mathcal{L}_{GaugeEW} = -\tfrac{1}{4} F_W{}^{a\mu\nu} F_W{}^a{}_{\mu\nu} - \tfrac{1}{4} F_B{}^{\mu\nu} F_{B\mu\nu} - \tfrac{1}{4} F_{B'}{}^{\mu\nu} F_{B'\mu\nu} + \mathcal{L}_{EW}{}^{ghost} \qquad (6.3)$$

and

$$\mathcal{L}_{GaugeC} = \mathcal{L}_{CCG} + \mathcal{L}_C{}^{ghost} + \mathcal{L}_{CC}{}^{ghost} \qquad (6.4)$$

The ElectroWeak gauge bosons $W_\mu{}^a$, B_μ and B'_μ field tensors are:

$$F_W{}^a{}_{\mu\nu} = \partial W^a{}_\mu/\partial x^\nu - \partial W^a{}_\nu/\partial x^\mu + g_2 f^{abc} W^b{}_\mu W^c{}_\nu \qquad (6.5)$$

$$F_{B\mu\nu} = \partial B_\mu/\partial x^\nu - \partial B_\nu/\partial x^\mu \qquad (6.6)$$
$$F_{B'\mu\nu} = \partial B'{}_\mu/\partial x^\nu - \partial B'{}_\nu/\partial x^\mu$$

$\mathcal{L}_{EW}{}^{ghost}$ is the Faddeev-Popov ghost terms for the ElectroWeak $W_\mu{}^a$ gauge bosons. The complexon color gluon lagrangian $\mathcal{L}_{CCG}$ is defined by

$$\mathcal{L}_{CCG} = -\tfrac{1}{4} F_{CC}{}^{a\mu\nu}(x) F_{CC}{}^a{}_{\mu\nu}(x) \qquad (6.7)$$

where

$$F_{CC}{}^a{}_{\mu\nu} = \partial/\partial X_c{}^\nu A_C{}^a{}_\mu - \partial/\partial X_c{}^\mu A_C{}^a{}_\nu + g f_{su(3)}{}^{abc} A_C{}^b{}_\mu A_C{}^c{}_\nu \qquad (6.8)$$

---

[63] The lagrangian below is much the same as that of Blaha (2011c) except for the change necessitated by the additional group SU(2)⊗U(1) in place of a U(1) part.

where $A_{C\ \nu}^{\ a}$ is the color gluon gauge field, g is the color coupling constant, and the $f_{su(3)}^{\ abc}$ are the SU(3) structure constants.

In addition $\mathcal{L}_C^{ghost}$ is the color SU(3) Faddeev-Popov ghost terms defined in appendix 19-A of Blaha (2011c) for the complexon Lorentz gauge and $\mathcal{L}_{CC}^{ghost}$ is the complexon color SU(3) constraint ghost terms defined through the Faddeev-Popov mechanism. The mass sector $\mathcal{L}_{Mass}$ is based on the Higgs Mechanism or the Dimensional Mass Mechanism (sections 21.3 and 21.4 of Blaha (2011c)) which creates the fermion and ElectroWeak vector boson masses, and generation mixing.

The lagrangian is supplemented with the following condition on all complexon fields $\Phi_{...}$:[64]

$$\nabla_r \cdot \nabla_i \Phi ... = 0 \tag{6.9}$$

Non-complexon fields Ω... in the left-handed formulation under consideration satisfy the subsidiary condition:

$$[\nabla_r \cdot \nabla_i - (\nabla_r^2 \nabla_i^2)^{1/2}]\Omega ... = 0 \tag{6.10}$$

which guarantees a complexon's real momentum is parallel to its imaginary momentum.

[64] These conditions implement eq. 5.50, the orthogonality of the real and imaginary parts of complexon 3-momentum.

Appendix 6-B. Postulates for a Derivation of The Complexon Standard Model

In Blaha (2011c) we showed that Logic did not have paradoxes and that Gödel's Undecidability Theorem was not a demonstration of the inconsistency of Logic but rather a fault in the specification of a correct subject for the predicate of the undecidable statement. We then went on to define a set of primitive terms and axioms, emulating Euclid's logical development of geometry, which would furnish a basis for the Standard Model in its extended form, The Complexon Standard Model. The hope was to show that there was an underlying logical theory based in space-time geometry that leads to The Standard Model – a theory that has been disparaged for over thirty-five years despite its long list of experimental successes.

In this appendix we will refine the set of postulates of Blaha (2011c) based on a better understanding of its space-time basis. In particular, the concept of the Reality group that we introduced in Blaha (2012a) is crucial to revealing the rationale for the (broken) symmetry group of The Standard Model.

In appendices A and B we carry the set of axioms to a yet lower level by relating particle physics to Asynchronous Logic and Computer Grammars at its most fundamental level – an approach that harkens back to the *monads* of Leibniz. It is somewhat ironic that computer scientists today seek to create quantum computers using single particles as logical memory units while our fundamental theory sees particles as fundamentally logical units with time as computer program execution – step by step – yet continuous; and particles in space as computer memory.

POSTULATES FOR A STANDARD MODEL THEORY

1. Any observer can define a set of four physical, real valued time and space coordinates called an inertial coordinate system in which the observer is at rest. The underlying Space-time has complex coordinates in general.

2. One can define a transformation that relates the inertial coordinate systems of two observers. The speed of light is the same in all inertial coordinate systems.

Conclusion: The transformation between coordinate systems must be a transformation in the complex Lorentz group.

3. Physical measurements of time or spatial coordinates are always real-valued numbers.

4. If a complex Lorentz transformation transforms the real-valued coordinates of an event to the coordinates of another coordinate system then, if the resulting coordinates are complex, a U(4) transformation exists that transforms the complex coordinates of the event to the real valued physical coordinates of the event. This transformation group is the Reality group.

5. The invariant interval or distance $d\tau$ in any coordinate system is defined by

$$d\tau^2 = g_{\mu\nu}dx^\mu dx^\nu \qquad (6\text{-B.1})$$

where $g_{\mu\nu}$ is a second rank tensor and the distances/time dx^μ are the physical coordinates of the interval.

6. The free dynamical equations of the four species of fermions can be generated by L(C) boosts from the Dirac equation for a fermion at rest where the set of allowed boosts $L_B(C)$ generate a 4-momentum in which the energy is real valued. All four types of species, including two tachyonic species, are realized in nature. The two quark species both have complex 3-momenta. One quark species has positive mass squared. The other quark species has negative mass squared.

7. Flat 4-dimensional space-time as well as a curved 4-dimensional space-time can be embedded in a 16-dimensional flat, complex space – the Flatverse. The transformations of the 4-dimensional Reality group U(4),

which has 16 generators, map to a Reality group in the Flatverse – the direct product group G_{16} = SU(3)⊗SU(2)⊗U(1)⊗SU(2)⊗U(1), which also has 16 generators.

Conclusion: The regular representations of G_{16}'s factors form a fully reduced 16 by 16 block matrix representation. One can form a covariant derivative using the 16 connections of G_{16}. This covariant derivative becomes a covariant derivative of the same form in four dimensions but uses the fundamental matrix representation for each G_{16} factor. Upon the introduction of fields with corresponding symmetries we can form dynamical equations and corresponding lagrangians. Second quantization leads to the Complexon Standard Model by 1.) requiring ElectroWeak doublets consisting of a non-tachyon field and a tachyon field, subsequently combined into quadruplets to include Dark Matter; 2) requiring the Hamiltonian be positive (thus favoring a left-handed theory); 3) following the conventional second quantization procedure.

8. Three (or possibly four) generations of fermions are introduced.

9. Lagrangian terms for Higgs particles (with generation mixing) and the gauge fields (connections) are added. All arguments of all fields are replaced with Two-Tier quantum coordinates. A free Two-Tier lagrangian dynamic term is also added. Two-Tier coordinates express the 4-dimensional coordinates of our universe in terms of Flatverse coordinates. Quantum fields that are arguments of Flatverse coordinates can be transformed to have quantum coordinate arguments.[65] Under the assumption that our universe is flat the Flatverse can be defined as 4-dimensional.

10. Using the lagrangian so obtained we create a path integral formulation and calculate quantities of interest.

The preceding list of postulates presumes many details of second quantization and other features. Yet it encapsulates the essence of The Complexon

---

[65] See chapters 4 and 5 of Blaha (2012b).

Standard Model with Dark Matter. In our defense we note that the paradigm of axiomatic theories, Euclidean geometry, has more than the five famous postulates embedded within its derivations of theorems. Hidden in the derivations are assumptions not made in the five postulates about the form of its geometric diagrams.

7. Origin of Higgs Particles' Masses in a Sister Universe within the Flatverse

7.1 Higgs Particles

Recently tentative experimental evidence[66] of Higgs particles with masses around 126 GeV has been found. Higgs particles are used to give masses to fermions and gauge boson fields as well as mixing angles in Standard Models of Elementary Particles.

In Blaha (2011c), and earlier books, we suggested a possible alternative to the Higgs mechanism based on an analogy with the Schwarzschild solution of General Relativity. A mass appears in this solution as a separation constant in the solution of the gravitational dynamic equations. This mass is then interpreted as the mass of a Black Hole. Similarly, we suggested a type of dynamic fermion and gauge boson dynamical equation in which particle masses appear as separation constants. This approach avoids the introduction of particles.

The apparent discovery of Higgs particles makes this alternate mechanism unnecessary at first glance. However the appearance of mass terms in the Higgs bosons' dynamic equations shows that, although the Higgs mechanism may generate masses for fundamental fermions and vector bosons it has not resolved the issue of the origin of mass and thus inertia.

It has "merely" pushed the question of the origin of mass one level further back so that it resides solely in the Higgs sector dynamic equations rather than having mass constants appearing for fermions and vector bosons independently. One could be content at this stage with placing the origin of masses solely in the Higgs sector of The Standard Model.

However one could also view this situation as opening a door into Reality only to find another deeper door as yet unopened. While there may be another solution to the origin of mass at this yet deeper level, the Dimensional Mass Mechanism of Blaha (2011c) resolves this problem by having terms with mass dimensions originate as separation constants as they do in the solution of the Schwarzschild dynamic equations.

---

[66] At the CERN LHC.

7.1 Generation of Higgs Particle Mass Terms

We will consider an example of the generation of Higgs particle mass terms for the case of complex scalar particle fields.[67] The realistic case of multiplets of Higgs particles is a direct generalization. The lagrangian density has the form

$$\mathcal{L}(\phi(x)) = \partial^{\mu}\phi\partial_{\mu}\phi^* - V(\phi, \phi^*) \quad (7.1)$$

where

$$V(\phi, \phi^*) = m^2\phi\phi^* + \beta(\phi\phi^*)^2 \quad (7.2)$$

If $\beta > 0$ and $m^2 < 0$ then spontaneous symmetry breaking occurs and the ϕ field acquires a constant part ϕ_0

$$\phi(x) = \phi_0 + \delta\phi(x) \quad (7.3)$$

with

$$\phi_0 = -m/\sqrt{\beta} \quad (7.4)$$

up to a phase factor. Note that a negative value for m^2 indicates the ϕ field is a tachyon field. The tachyonic nature is transmuted to be a constant throughout space and time by the quartic interaction term.

We note further that β is a dimensionless constant. Consequently the mass term is the only "dimensioned" constant in the lagrangian and the resulting dynamical equations.

We will now show that the mass term can originate through a separation of equations IF the true lagrangian is the sum of pieces from different universes. We call this the Dimensional Mass Mechanism. (See Blaha (2011c).)

We assume n Higgs bosons that are functions of our space-time coordinates labeled x and four space-time coordinates labeled y in another space-time (universe).[68]

The familiar Higgs boson field equations have a quartic interaction if we ignore the other universe and its contribution of a mass term. Let us define

---

[67] Chapter 1 of Blaha (2012b) contains this discussion and example, which we place here for the sake of completeness.

[68] See Blaha (2011c) for a more detailed discussion of the "other" space-time.

$$B^a(x) = \Box\phi^a + \beta_a\phi^{a4} = 0 \quad (7.5)$$

Following a parallel development to the fermion case of Blaha (2011c) we consider a coupled universes' generalization of eq. 7.5:

$$B^a(x, y) \equiv B^a(x)_{U1}\, \phi^a(y)_{U2} + \phi^a(x)_{U1} B^a(y)_{U2} = 0 \quad (7.6)$$

where U1 signifies our universe and U2 signifies the other "sister" universe.

This equation results from the lagrangian density term

$$V_a(x, y)\, B^a(x, y) \quad (7.7)$$

where $V_a(x, y)$ is a subsidiary field. Variation with respect to $V_a(x, y)$ yields eq. 7.6. Variation with respect to $\phi^a(x)_{U1}$ and $\phi^a(y)_{U2}$ yield the other dynamical equations. The linearity in $V_a(x, y)$ in these dynamical equations imply $V_a(x, y)$ is separable and the product of two boson fields:

$$V_a(x, y) = V_{1a}(x) V_{2a}(y) \quad (7.8)$$

Dividing eq. 7.6 by $\phi^a(x)_{U1}\phi^a(y)_{U2}$ yields a separable equation with the result:

$$B^a(x)_{U1}/\phi^a(x)_{U1} = -m_a^2 \quad (7.9)$$

and

$$B^a(y)_{U2}/\phi^a(y)_{U2} = m_a^2 \quad (7.10)$$

where the right side of each equation is a constant mass squared (required on dimensional grounds).

Consequently a Higgs boson field equation in our universe acquires mass terms

$$\Box\phi^a + m_a^2\phi^{a2} + \beta_a\phi^{a4} = 0 \quad (7.11)$$

and thus provides an origin for the Higgs mass term based on the separation of variables. The existence of another universe is the price for a deeper understanding the origin of mass.

7.2 Does the Flatverse Contain Another Universe?

The Flatverse contains our complex 4-dimensional universe. Our universe can be mapped to 8 complex dimensions of the Flatverse since the metric tensor of our complex universe is complex in general and has 16 real valued parameters (or 8 complex parameters) due to eqns. 4.1 and 4.1a.[69] These 16 parameters can, in turn, be mapped into an 8 complex dimension subspace of the Flatverse.

The other 8 complex dimensions of the Flatverse form the basis of another sister universe that we can assume contains a complex 4-dimensional universe (perhaps not unlike our own universe in some respects). In appendices A and B we show that elementary particle physics can be viewed as a consequence of Asynchronous Logic and has two universes for reasons given there.

So the discussion in the previous section which relies on a second universe is supported by our Flatverse and is justified by a deeper basis constructed on Asynchronous Logic.

If this second universe exists, and the Dimensional Mass Mechanism uses it as shown in the preceding section, then the Higgs particles provide a "gateway" of sorts between the universes. Since the Higgs' masses squared in the sister universe have opposite sign to the Higgs masses squared in our universe, spontaneous symmetry breaking doesn't occur in the same way as it does in our universe. The other universe's equivalent of the ElectroWeak gauge particles and Dark gauge particles are massless. Thus their Weak Interactions are long range interactions. Normal fermions and tachyonic fermions would appear to have opposite roles in the other universe compared to the fermions in our universe. The other universe would thus be very different physically although the same physical laws would hold.[70]

7.3 Origin of Fermion Generations

We believe that the origin of the fermion generations is due to the existence of a second sister universe – the same universe that lead to the Higgs

[69] The complex 4-dimension metric tensor has 32 real parameters (16 complex parameters) that are reduced by eq. 4.1a to 16 real parameters or 8 complex parameters.

[70] We defer further detailed consideration of the sister universe to a later time. The next section has a preliminary discussion related to the fermion generations problem.

mass terms as described in section 7.1. In this section we will describe the source of fermion generations.[71] We will call our universe U1 and the sister universe U2.

A free Dirac-like dynamical equation is linear in coordinate derivatives and has the general form

$$(\gamma^{\mu}\mathcal{D}_{\mu}(x) - m)\psi(x) = 0 \qquad (7.12)$$

In the case of the direct product of the universes, denoted U1⊗U2, we define the U1 4-vector coordinates as x^{ν} and the U2 4-vector coordinates as y^{μ} for the purposes of the discussions in this section and the following sections of this chapter. Each species[72] composite particle[73] Dirac-like equation becomes a 4 generation first order Dirac-like equation of the form[74]

$$[(D_{U1} + V_{U1})\otimes I_{U2} + I_{U1}\otimes(D_{U2} + V_{U2}) + V_{U1,U2}]\psi(x, y) = 0 \qquad (7.13)$$

where I represents the identity matrix, and where V_{U1}, V_{U2}, and $V_{U1,U2}$ are interaction/potential terms. Explicitly with indices displayed eq 7.13 is

$$(\gamma^{\nu}{}_{ab}\mathcal{D}_{U1\nu}(x) + V_{U1ab}(x))\delta_{cd} + (\gamma^{\mu}{}_{cd}\mathcal{D}_{U2\mu}(y) + V_{U2cd}(y))\delta_{ab} + V_{U1,U2ab,cd}(x, y))\psi_{bd}(x, y) = 0 \qquad (7.14)$$

The index d will become the generation index. The index b is the spinor index of our space-time. Consequently there are four equations of the form of eq. 7.12 – one equation for each fermion generation.[75] The 4 solutions $\psi_{bd}(x, y)$ for d = 1, 2,

---

[71] This proposal differs from that of chapter 21 of Blaha (2011c) which was written before Higgs particles were found experimentally.

[72] There are also the 8 species of fermions to consider: charged leptons, uncharged leptons, Dark charged Dark leptons, Dark neutral Dark leptons, up-type quarks, down-type quarks, and Dark up-type quarks, and Dark down-type quarks.

[73] Composite in the sense that in the simplest quantum mechanics model of the hydrogen atom the wave function is the product of the electron wave function and the proton nucleus wave function. The proton wave function factor is discarded from the composite wave function except for its mass and spin since the proton is treated as a very massive (immobile) particle. The mass appears in the electron energy level expression and the electron energy levels are split due to electron spin-orbit interaction with the proton spin. The analogy with the present case is clear. The sister universe dynamics causes the generations and their mass spectrum splitting. Yet being a separate universe it has no other directly observable features in particle interactions in our universe. It does have other indirect but important effects (chapter 8) that support its physical existence.

[74] The form is largely determined by the requirement that Dirac-like equations be first order equations.

[75] A subsidiary condition could reduce the number of generations to three generations. However experimental evidence for a fourth generation has started to appear.

3, 4 correspond to the four generations of a species. Taking account of Dark Matter species there are 4 generations of the 8 species and thus there will be 8 mass matrices when we separate the equations.

Eq. 7.14 is a separable equation if $V_{U1,U2}$ supports separability.[76] We assume it is separable. We can express the wave function of a species with generations labeled by the index d as[77]

$$\psi_{bd}(x, y) = \psi_{bd}(x)\psi'_d(y) \tag{7.15}$$

An index for each of the 8 species is not displayed for simplicity and also because eq. 7.14 does not mix species. If $V_{U1,U2ab,cd}(x, y)$ is a Higgs term that is effectively of the form[78]

$$V_{U1,U2ab,cd} \rightarrow \mathcal{M}_{cd}\delta_{ab}$$

we can introduce a matrix of separation constants $\mathcal{M}_{cd}$. Then we find that eq. 7.14 is solved using the solutions of the separated equations:

U2 universe: $$[\gamma'^{\mu}{}_{cd}\mathcal{D}_{U2\mu}(y) + V_{U2cd}(x)]\psi'_d(y) = 0 \tag{7.16}$$

U1 universe: $$(\gamma^{\nu}{}_{ab}\mathcal{D}_{U1\nu}(x)\delta_{cd} + V_{U1ab}(x) - \mathcal{M}_{cd}\delta_{ab})\psi_{bd}(x) = 0 \tag{7.17}$$

Note $\mathcal{M}_{cd}\delta_{ab}$ is the mass terms of the generations with a mixing matrix.[79] It has a similar flavor to the simplest solutions of the hydrogen atom in quantum mechanics where the nucleus (assumed to be spin ½ and have infinite mass) determines energy levels and through spin interactions splits electron orbital energies through spin-orbit interactions.

Thus we find a mechanism that generates the Higgs masses and in turn the fermion generations and their mass matrix through the introduction of a sister universe U2.

---

[76] The aforementioned hydrogen atom has a separable potential that is achieved by the use of the relative distance of the electron from the proton.

[77] The reader who is concerned about treating the wave function classically rather than as a second quantized field should remember that the manipulations that we are performing can be redone in a path integral formalism in which the manipulation of fields as classical fields is perfectly acceptable.

[78] Note a and b are spinor indices in our U1 universe.

[79] The overall form is similar to the Cabibbo-Kobayashi-Maskawa matrix.

The fermion fields of each species implied by eq. 7.17 have the form:

$$\psi(x) = \begin{bmatrix} \psi_{d=1} \\ \dots \\ \psi_{d=4} \end{bmatrix} \tag{7.18}$$

with each ψ_d a four component coordinate space Dirac spinor.[80] More details of fermion generations and mass matrices appear in Blaha (2011c).

[80] We note that $\psi'_d(y)$'s only role is to support generations and the construction of the mass matrices.

8. Evidence for the Flatverse and for a Sister Universe to Our Universe within It

8.1 Large Scale Structure of of the Flatverse

In chapter 7 we introduced a sister universe within the Flatverse that enabled us to explain the origin of the Higgs particles' masses and fermion generations. The Higgs particles use the mass terms so generated in their part of the Higgs sector, plus quartic interaction terms, to initiate a spontaneous breakdown that gives masses to the fundamental fermions and gauge bosons in our extended Complexon Standard Model.

Our universe treated as a complex[81] 4-dimensional space-time is embedded in a 16 dimension complex space. Within the 16 complex dimension space there is "room" for another complex 4 dimensional space-time that we call the sister universe. In chapter 7 we show the masses of Higgs particles can be generated from the combined lagrangian terms for Higgs particles in our universe and our sister universe as separation constants in the Higgs dynamical equations.

Thus we view the Flatverse as composed of two universes that differ in their internal details.

If we do not accept this interpretation for the origin of the Higgs mass terms then we will not be able to understand the origin of the Higgs masses and consequently the masses of fundamental particles in our universe. The Higgs Mechanism will then simply push the mystery of the source of masses to a deeper level where the same question of the origin of mass will reappear.

We view the separation of variables approach as the only currently viable approach to the understanding of the origin of mass.

---

[81] The complex coordinates of our space-time are mapped to real valued physical coordinates by the Reality group.

8.2 Mach vs. Newton vs. Einstein on the Origin of Inertial Reference Frames

It is fitting, that the use of the Flatverse with an embedded sister universe to obtain the origin of mass, should also resolve the issue of the origin of inertial reference frames – a controversial issue since the work of Mach over 150 years ago – an issue considered by the greats of General Relativity such as Einstein. The major schools of thought on the definition of inertial reference frames are linked to great names in Physics and we will so identify them realizing that major contributions were also made by many other physicists.[82]

8.2.1 Newton's Definition of Inertial Frames

Newton posited the existence of an absolute reference frame.[83] He then defined inertial frames as those frames that are at rest with respect to the absolute frame or in a constant state of motion with respect to the absolute frame. This definition was the accepted definition until the analysis of Mach in the 19th century.

8.2.2 Mach's Definition of Inertial Frames

In the mid-19th century Mach suggested that the definition of reference frames be based on the total mass distribution of the universe. This is described as "with reference to the fixed stars" which were thought to provide an equivalent to Newton's absolute frame although there were issues with non-uniformities in large local mass distributions that might compromise the definition of inertial frames in some regions.

8.2.3 Einstein/General Relativity Definition of Inertial Frames

Einstein proposed an equivalence principle to explain the origin of inertial frames. The local gravitational field at any point determines a free falling inertial frame. All dynamical equations in this inertial frame are not affected by the existence of nearby masses. Thus the Equivalence Principle has a Machian flavor since all masses determine the local gravitational field including the "fixed stars." But one could consider the set of all such inertial frames as all related by

---

[82] See Blaha (2004) for a survey of the thoughts of some major General Relativists on inertial frames and absolute frames.

[83] He attributed the existence of this frame to God.

transformations. Then one could single out one frame as aNewtonian absolute frame.

8.2.4 Flatverse Mach's Definition of Inertial Frames

We have proposed the existence of a flat 16 dimension complex space based on a number of considerations in previous chapters and books, and later in this chapter. Our universe is a surface within the Flatverse. We can then define inertial frames in our universe as those frames that are at rest with respect to the Flatverse or in a constant state of motion with respect to the Flatverse. Thus we have a well-defined inertial frame definition. The reader will note that it is similar to Newton's definition but it does not have a theological rationale.[84] Newton's absolute frame had four real valued dimensions but the Flatverse has 16 complex dimensions.

8.3 Rationale for Connections – Dark and Normal ElectroWeak Gauge Fields and SU(3) Gluon Fields

The origin of the known Standard Model gauge fields *was* unknown. As a result of the studies here and previous books by the author it has become clear that SU(3)⊗SU(2)⊗U(1)⊗SU(2)⊗U(1) is the correct Standard Model symmetry on geometrical grounds. It has also become clear that it is based on the connections of the Reality group in complex 16 dimension space and in our 4 dimension space-time. Chapter 4 presents the geometric basis as do our books written in 2012.

It is also clear that the Flatverse is a crucial part of our understanding of the Standard Model gauge fields (connections). Thus the symmetry of the extended Complexon Standard Model, of which the SU(3)⊗SU(2)⊗U(1) part is experimentally verified, and the additional SU(2)⊗U(1) is a clear candidate for the Dark Matter sector, significantly supports the existence of the Flatverse.

---

[84] In 1964 (or thereabouts) this author wrote a paper (now lost) for a Philosophy course that compared the concepts of nature of early Presocratic Greek philosophers with the prior, and then prevailing, religious concepts of nature. In my paper I showed that the logic of the Presocratic philosophers was similar to that of the "theologians" of the time. The difference was that the philosophers substituted natural, non-religious concepts for religious concepts. The beginnings of science then can be viewed as a demythologizing of concepts but keeping the same logical framework. In the present case we replace Newton's absolute space based on theology with an impersonal physical concept – the Flatverse. See G. S. Kirk and J. E. Raven, *The Presocratic Philosophers*, (Cambridge University Press, Cambridge, UK, 1962) for information on Presocratic philosophy.

8.4 Flatverse Explanation for a Finite Standard Model - Renormalization

The Two-Tier mechanism for making The Complexon Standard Model finite to all orders in perturbation theory can be understood in terms of a transition from c-number coordinates in the Flatverse to q-number coordinates in our universe, a surface within the Flatverse. See section 4.9 and chapter 4 of Blaha (2012b). Two-Tier renormalization is discussed in detail in Blaha (2005a) as well as earlier books by the author.

In chapter 9 we will see that the q-number part of quantum coordinates plays a major role in producing a finite Big Bang (no divergences) and an explosive inflationary expansion.

8.5 Origin of the Fermion Generations

Chapter 7 explains the mystery of the origin of the Higgs particles' mass terms as separation constants[85] of a larger dynamical equation for Higgs particles that spans the Flatverse. As pointed out in chapter 7 and in Blaha (2012b) an understanding of the origin of mass and inertia requires more than the Higgs Mechanism since the origin of the mass terms in the Higgs dynamic equations is not specifiable without using separation constants. Any other approach simply puts off the resolution of the origin of mass and inertia.

The key feature of the separation constants approach is the existence of a sister universe of 8 complex dimensions in the Flatverse in addition to our universe (which is embedded in the other 8 complex dimension surface within the Flatverse.) Thus we find the Flatverse consists of two 8 complex dimension surfaces within the Flatverse: our universe and our sister universe.

The Flatverse is then essential to resolve the mystery of mass and inertia.

8.6 Genesis: Universe—Anti-Universe Creation?

Many universe theories are often considered in one form or another by theoretical physicists. In our theory two universes are well motivated by the features and requirements of The Standard Model. The Flatverse makes The Standard Model understandable geometrically.

---

[85] Blaha (2012b) shows that the mass term in the Schwarzschild solution of General Relativity is a separation constant. We suggest Nature uses the same approach for Higgs mass terms.

However the introduction of a sister universe in Blaha (2012b) and this book raises the question: I may understand a Big Bang wherein one universe is created. But how do you create a sister universe at the same time? While one could concoct schemes where this scenario would happen, we cannot honestly propose a detailed mechanism with any certainty at this point.

But there is one qualitative, experimental hint of a sister universe. It is well known that there is a vast predominance of particles over antiparticles in our universe. This raises serious questions with respect to discrete and gauge symmetries.

These issues could perhaps be resolved if the universes were generated by a quantum fluctuation. A Flatverse fluctuation[86] could create a universe–anti-universe pair with the contents of each universe complementary—one universe containing primarily particles, and the other universe containing primarily anti-particles respectively. Exploring this promising possibility is beyond the scope of this book.

8.7 A Basis for the Big Bang that Generated Our Universe

The Two-Tier coordinates that we introduce within the framework of the Flatverse appear to be able to 1) keep the origin of universe – the Big Bang – from being singular through what may be called quantum smearing; and 2) can explain the inflationary growth of the universe. In chapter 9 we will develop this theory of the Big Bang and expansion. It is similar to the Big Bang theory we developed in Blaha (2004). The existence of the Flatverse creates a framework for this Quantum Big Bang theory.

8.8 Asynchronous Logic adapted to Particle Theory suggests Two Universes

In appendix A (and in earlier books such as Blaha (2011c)) we suggest that a strong analogy exists between Asynchronous Logic (the logic relevant to Very Large Scale Integration (VLSI) computer chip circuit designs) and elementary

---

[86] A Flatverse capable of generating fluctuations seems to require a dynamic vacuum, and quantum fields. This possibility opens the door to many questions: What dynamic quantum fields? What about gravitation and quantum gravitation? What are the features of an empty Flatverse that would exist before the fluctuation? Were there prior fluctuations? Will there be more fluctuations that might perhaps expunge the fluctuation that we call our universe? Can other universes exist that do not interact with our universe and are superposed with our universe partly occupying our universe's space without our knowledge? And so on.

particle theory. This foundation if pushed leads to two four dimensional spaces that we interpret as our universe and our sister universe. These spaces, when combined, form a sixteen dimensional space that we identify with the Flatverse.

Appendix A provides details of the need for two 4-dimensional spaces—the basis of the Flatverse.

9. Dark Energy and The Big Bang

Nature abhors a singularity.

Much of this chapter first appeared in Blaha (2004). In the light of our progress since then we now provide a somewhat revised version that is similar to the 2004 version for the most part.

The current state of our knowledge of the evolution of the universe has now been extended back in time to about 350,000 years after the Big Bang through recent astrophysical research. While this progress is encouraging we still face major issues: the nature of Dark Matter (hopefully resolved in chapter 1), the nature and origin of Dark Energy (hopefully resolved in this chapter), and the events of those critical years before the 350,000 year point that we have now apparently reached experimentally. Those early years and the Big Bang itself remain mysteries. This situation is especially critical since the early years of the universe apparently contain an uncertain beginning and explosive growth.

In this chapter we will attempt to understand that unknown period in the neighborhood of $t = 0$ where quantum effects we believe play a major role. We will suggest that the inflationary growth of the universe, which is attributed to an unknown "particle" called an inflaton, actually is caused by the energy of the q-number part of coordinates – a quantum field denoted $Y^{\mu}(x)$ in earlier chapters.

The inflaton is the $Y^{\mu}(x)$ quantum field. We will see $Y^{\mu}(x)$ makes the Big Bang finite—no singularity; and frees The Standard Model and Quantum Gravity[87] of infinities. $Y^{\mu}(x)$ has a remarkable triple role in our view – to eliminate the Big Bang singularity, to generate the explosive growth of the universe, and to remove infinities from The Standard Model and Quantum Gravity.

Since the $Y^{\mu}(x)$ quantum gauge field is a free field (neglecting gravity) the initial state of the universe can be permeated with quanta of this field as well as quanta generated by particles. The total energy of the free $Y^{\mu}(x)$ field within the universe is then the energy that we call Dark Energy—energy that can only influence the universe through its gravitational effects.

---

[87] See Blaha (2011c) and Blaha (2005a) for the removal of infinities in Quantum Gravity.

9.1 The State of the Universe at $t = 0$

If we extrapolate the currently popular Standard Cosmological Models (with or without inflations) back to the Big Bang t = 0, we find a universe beginning as a single "mathematical point" with infinite mass density and infinite temperature. The Robertson-Walker metric scale factor a(t), which is a solution of the Einstein equation

$$\dot{a}^2 - 8\pi G\rho a^2/3 = -k \tag{9.1}$$

typically is solved for a perfect fluid under the assumption of a matter-dominated or a radiation-dominated phase of the universe. If we assume the universe is matter-dominated, then the energy density is

$$\text{Matter-Dominated:} \qquad \rho = \rho_0/a(t)^3 \tag{9.2}$$

Under the alternate assumption that the universe is radiation-dominated we have

$$\text{Radiation-Dominated:} \qquad \rho = \rho_0/a(t)^4 \tag{9.3}$$

With either assumption we find that the scale factor behaves as

$$a(t) \propto t^n \tag{9.4}$$

where $0 < n < 1$. Thus $a(0) = 0$ and the universe reduces to a point with infinite density (eqns. 9.2 and 9.3), and with infinite temperature since

$$T \propto a^{-1}(t) \tag{9.5}$$

There are evidently grave difficulties in extrapolating the Standard Cosmological Model, or its current variants, to t = 0. The difficulty is compounded by the inherently quantum mechanical aspects that are normally associated with gravitation at ultra-small distances.

Currently, the only viable complete theory of Quantum Gravity is the Two-Tier Quantum Gravity of Blaha (2003) and (2005a). Blaha's type of quantum field theory has the interesting feature that all forces (particle propagators) become zero at very short distances (presumably much less than the Planck mass). Thus a point universe could have an infinite density of essentially "non-interacting" matter as a quasi-stable state. Furthermore if one uses the *generalized* Robertson-Walker metric as described in Blaha (2004) one finds a classical scale factor of the form:

$$A(t, \check{r}) = a(t)b(\check{r}) \tag{9.6}$$

where a(t) satisfies eq. 9.1.

Since quantum effects can be expected to play a role near t = 0 (the Big Bang) it is possible that the expectation value of a quantum scale factor operator, taking account of quantum effects, could have the form

$$<A(t, \check{r})> = <a(t)><b(\check{r}, t)> \tag{9.7}$$

where

$$<b(\check{r}, t)> \rightarrow \beta(\check{r})/<a(t)> \tag{9.8}$$

as $t \rightarrow 0$ so that the zero of <a(t)> might be cancelled with the result

$$<A(t, \check{r})> \rightarrow \beta(\check{r}) \neq 0 \tag{9.9}$$

Quantum effects would thus eliminate the singularities at t = 0. A quantized version of the generalized Robertson-Walker model[88] opens the possibility of a universe with a finite size, density, and temperature at the time of the Big Bang.

With that possibility in mind, we will use Blaha's (2004) Two-Tier theory to develop an amended version of his quantum model of the universe in the neighborhood of t = 0. Starting from eq. 8.7.1 of that book which is eq. 9-A.7.1 in the following appendix:

$$A(t, \check{r}) = a(t)b(\check{r}) = 2\alpha k^{-1/2}a(t)[1 + \alpha^2\, \check{r}^2]^{-1} \tag{8.7.1}$$

and introducing a Two-Tier variable Y (as in chapter 7 of Blaha (2004)) with the identification[89]

---

[88] This model appears in Blaha (2004).

[89] α is a constant and not the fine structure constant.

$$\check{r} \equiv M_cX = M_c(y + iY/M_c^{2})$$

we see

$$b(\check{r}) = b(y, t) = 2\alpha k^{-½}[1 + \alpha^2(M_cy + iY/M_c)^2]^{-1} \tag{9.10}$$

If

$$Y = -M_c[a_1(y, t) - a_2(y, t)a(t)]^{½}/\alpha + iM_c^{2}a_3(y, t) \tag{9.11}$$

and if, as t → 0,

$$a_1(y, t) \rightarrow a_1(y, 0) = 1 \tag{9.12}$$

$$a_2(y, t) \rightarrow a_2(y, 0) \neq 0 \tag{9.13}$$

$$a_3(y, t) \rightarrow a_3(y, 0) = y \tag{9.14}$$

then we find

$$b(y, t) \rightarrow 2\alpha k^{-½}/[a_2(y, 0)a(t)] \tag{9.15}$$

and

$$A(t, \check{r}) = a(t)b(\check{r}) = 2\alpha k^{-½}/a_2(y, 0) + a(t)\beta_1(y) + a^2(t)\beta_2(y) + \ldots \tag{9.16}$$

$$\rightarrow 2\alpha k^{-½}/a_2(y, 0) \qquad \text{as} \quad t \rightarrow 0$$

where we omit symbols indicating expectation values for the sake of clarity. Thus, under these circumstances, space does not collapse to a point and the density and temperature—as well as other parameters of interest—are finite; *and the features of the Standard Cosmological Model at larger times are still valid.*

In our model we will make the following assumptions about the universe near t = 0:

1. The particles in the universe consist of fundamental elementary particles – gravitons, photons, electrons, neutrinos, quarks, gluons and so on – and their corresponding anti-particles.

2. The particles are described by Two-Tier quantum field theory. In this type of quantum field theory all particle interactions become

negligible at very short distances (as described in chapter 7 of Blaha (2004)) and so the forces between particles may be neglected near t = 0 when the universe is immensely *small*.

3. The energy of the universe can be viewed as consisting of particles – bosons and fermions – each species having blackbody energy distributions since the universe is the best of all possible black bodies.

4. The enormous energy of the universe even if confined to a small region makes the classical Einstein equations a good approximation *due to its macroscopic nature* with one proviso (item 5). Therefore we assume the Generalized Robertson-Walker metric of Blaha (2004).

5. In the neighborhood of t = 0 when the universe is effectively confined to a region whose scale is set by the Planck mass or smaller the quantum nature of the Two-Tier coordinate X^μ becomes significant. In particular the Y^μ field causes a profound change in the behavior of the scale factor A(t, r) as $t \rightarrow 0$. (Note: $X^\mu(y) = y^\mu + iY^\mu(y)/M^2$ defines the quantum coordinates where y^μ is a c-number coordinate and Y^μ a free q-number field similar to the electromagnetic field.)

6. The Y^μ quanta are assumed to have a black body spectrum[90] – just like elementary particles – reflecting their continuous emission and absorption by gravitons and other elementary particles. The Y^μ blackbody spectrum is implemented via a coherent state. Effectively the coherent state opens a small "bubble" into complex space-time changing the dynamics of the universe at t = 0.

7. We will calculate the expectation value of the quantum field operator Y^μ in a closed Robertson-Walker space. In principle we must use the generalized Robertson-Walker metric since the scale factor will depend on both r and t through its dependence on the expectation value of Y^μ.

---

[90] The only reasonable choice for the spectrum is a black body spectrum given the confinement of the field to the ultimate black body – the universe.

9.2 Two-Tier Quantum Model for the Beginning of the Universe

In our approach in this, and the following, sections we will follow a modest program using the known theoretical foundations of elementary particle physics: the Standard Model unified with Quantum Gravity in a Two-Tier quantum field theoretic framework. We will supplement this framework with natural assumptions about the initial conditions of the universe in order to develop a theory describing the evolution of the universe from its initial state.

9.2.1 Einstein Equations Near $t = 0$

There is no physical reason to believe that the universe at the beginning of time, t = 0, was a mathematical point of infinite temperature and density since the extrapolation of the scale factor of the Standard Cosmological Model to t = 0 is unwarranted for many reasons including quantum considerations.

Ideally we would use the Quantum Theory of gravity to establish the physical theory of the universe near t = 0. However a quantum calculation of the global structure of the universe near t = 0 is not feasible. In view of this situation we must find an approximation that captures the physics of the universe near the Big Bang. One approach is based on the macroscopic energy of the early universe. One can expect that a classical gravitation model theory with appropriate quantum corrections may be a reasonable approximation to the early state of the universe. After all, macroscopic bodies are described by classical physics in general. And the universe is a macroscopic body by virtue of its content at the point of the Big Bang despite its small size. Therefore we can assume that we may start with a classical gravitation model and then introduce quantum corrections.

The natural first choices – based on symmetry considerations – are a Robertson-Walker model and a generalized Robertson-Walker model of the type described in chapter 8 of Blaha (2004) and appendix 9-A following. The quantum part that we will shortly introduce will require us to use a generalized Robertson-Walker model since the quantum corrections reduce the symmetry to a maximally symmetric *two-dimensional* subspace within a four-dimensional space-time. *The quantum part eliminates the equivalence between the classical Robertson-Walker and generalized Robertson-Walker models* that was described in section 8.7 of Blaha (2004). See appendix 9-A.

Therefore we begin with the classical c-number equation for the invariant interval defined by

$$d\tau^2 = dt^2 - A^2(t, \check{r})[d\check{r}^2 + \check{r}^2(d\theta^2 + \sin^2\theta\, d\varphi^2)] \tag{9.17}$$

where

$$A(t, \check{r}) = a(t)b(\check{r}) = 2\alpha k^{-1/2} a(t)[1 + \alpha^2\, \check{r}^2]^{-1} \tag{9.18}$$

and a(t) is the solution of the Einstein equation:

$$\dot{a}^2 - 8\pi G\rho a^2/3 = -k \tag{9.19}$$

Next we introduce quantum coordinates

$$X^\mu = y^\mu + i\, Y^\mu(y)/M_c^{\,2} \tag{9.20}$$

We choose the same transverse gauge for Y^μ as we did in chapter 7 of Blaha (2004):

$$\partial Y^i/\partial y^i = 0 \tag{9.21}$$

$$Y^0 = 0 \tag{9.22}$$

As a result we make the identification (definition of coordinates)

$$X^0 = y^0 \equiv t \tag{9.23}$$
$$X^j = y^j + i\, Y^j(y)/M_c^{\,2} \equiv M_c^{-1}\check{r}^j \tag{9.24}$$

The mass factor on the right side of the ≡ sign eq. 9.20 is required on dimensional grounds if y is to have the usual dimension of length (inverse mass). As a result since $\check{r} \in [0, 1]$ by Blaha (2004) eq. 9.24 implies

$$y = |\mathbf{y}| \in [0, M_c^{-1}] \tag{9.25}$$

There are two constants with the dimension of mass to a power: k and M_c. The constant k determines the curvature of space – a large-scale feature of Robertson-Walker models. The constant M_c is related to the very short distance behavior of the theory – high energy phenomena with energies of the order of the Planck mass or larger, and, as we will see, the origin of the universe – a short distance, high energy phenomena as well. Therefore we have also chosen to use M_c on the right side of eq. 9.24.

Since X^0 is a c-number and since the density $\rho(t)$ is a large c-number to very good approximation we will assume a(t) is the c-number solution of the classical c-number eq. 9.19 as t → 0. Further we assume that quantum effects appear solely through b(ř). We assume that the q-number equivalent of b(ř) is $b(M_cX)$. The function $b(M_cX)$ satisfies the functional equation:[91]

$$k + (M_c^2Xb^2)^{-1}\,\partial(Xb'/b)/\partial X = 0 \tag{9.26}$$

where

$$b' = \partial b/\partial X \tag{9.27}$$

and $X = (\vec{X}\cdot\vec{X})^{1/2}$. The formal solution of eq. 9.26 has the same functional form as the c-number solution b(ř) in eq. 9.18. Therefore

$$b(M_cX) = :2\alpha k^{-1/2}[1 + \alpha^2M_c^2X^2]^{-1}: \tag{9.28}$$

where we have specified normal ordering with : ... : to avoid trivial divergences.

Since eq. 9.28 is a q-number expression we must find the scale factor as the expectation value of A(t, X) for a suitable state. We note that, at this point, the invariant interval is an operator expression of the form:

$$d\tau^2 = dt^2 - B^2(t, X)[dX^2 + X^2(d\theta^2 + \sin^2\theta\, d\varphi^2)] \tag{9.29}$$

where

$$B(t, X) = M_cA(t, M_cX) = a(t)b_M(X) \tag{9.30}$$

and

$$b_M(X) = M_cb(M_cX) = :2\alpha M_ck^{-1/2}[1 + \alpha^2M_c^2X^2]^{-1}: \tag{9.31}$$

Thus the expectation value of $b_M(X)$ also must be calculated in order to determine the invariant interval's expectation value.

9.2.2 Y Black-Body Coherent States

The Y quanta are continuously being emitted and absorbed by the particles in the primeval universe. As such, they may be expected to have a

---

[91] See eq. 9-A.5.3 in appendix 9-A.

blackbody energy spectrum that is similar to that of the particles from which they derive their existence. In particular one expects the temperature T associated with their blackbody energy distribution to be the same as that of the "real" particles in the universe. After all, the universe is a black body.

Thus the blackbody energy of Y-quanta as a function of frequency ν per unit volume per unit frequency is assumed to be:

$$u_\nu = 8\pi hc^{-2}\nu^3\,[e^{h\nu/\kappa T} - 1]^{-1} \tag{9.32}$$

where c is the speed of light, h is Planck's constant, and κ is Boltzmann's constant. At this point we adopt units in which c = 1 and $\hbar = h/2\pi = 1$.

The Hamiltonian for the Y field has a form that is familiar from electrodynamics

$$H = \int d^3y\, \mathscr{H}_Y(y) = \tfrac{1}{2}\int d^3y\, {:}E_Y^2 + B_Y^2{:} = \int d^3p\, \omega \sum_\lambda a^\dagger(p,\lambda)a(p,\lambda) \tag{9.33}$$

where E_Y and B_Y are the "electric" and "magnetic" fields of the Y field, $\omega = p^0 = |\vec{p}| = 2\pi\nu$ is the energy (in our units), and λ labels the polarization. Note that we are using "infinite volume" continuum quantization formulation.

We now define coherent Y field bra and ket states that yield a spherically symmetric blackbody distribution as the eigenvalue of the Hamiltonian H:

$$|BB, T\rangle = N\ \exp[\int d^3p\, f(\omega,T)\sum_\lambda a^\dagger(p, \lambda)]\,|0\rangle \tag{9.34}$$

$$\langle BB, T| = N^*\langle 0|\exp[\int d^3p\, f^*(\omega,T)\sum_\lambda a(p, \lambda)] \tag{9.35}$$

where $\omega = |\vec{p}|$, and where N is a normalization factor. The expectation value of H is

$$\langle BB, T|H|BB, T\rangle = \int d^3p\, 2\omega|f(\omega,T)|^2 \tag{9.36}$$

$$= \int d\omega\, 8\pi\omega^3|f(\omega,T)|^2 \tag{9.37}$$

$$= \int d\nu\, 16\pi^2\omega^3|f(\omega,T)|^2 \tag{9.38}$$

where $\omega = 2\pi\nu$ in our units ($c = 1$, $\hbar = 1$), and where the factor of two in eq. 9.36 is the number of polarizations.

The expectation value (eigenvalue) of the energy per unit frequency is

$$H_\nu = 16\pi^2\omega^3 |f(\omega,T)|^2 \tag{9.39}$$

We relate H_ν to the blackbody energy *per unit volume* per unit frequency u_ν using

$$H_\nu = u_\nu(2\pi/\omega)^3 \tag{9.40}$$

where the factor of $(2\pi/\omega)^3$ makes the right side of eq. 9.40 the blackbody energy per unit frequency in the continuum case of a quantum field in a space of infinite volume. Thus we find

$$f(\omega,T) = \omega^{-3/2}[e^{\omega/\kappa T} - 1]^{-1/2} \tag{9.41}$$

with the phase of $f(\omega,T)$ set to zero.

9.2.3 Expectation Value of Y in Coherent States

As a preliminary to the evaluation of the operator scale factor in eq. 9.31 we will evaluate the expectation value of powers of the Y field between black body coherent states defined by eqns. 9.34-9.35. We will then determine the expectation value of $b_M(X)$ in combination with $a(t)$ to obtain the behavior of the overall scale factor near $t = 0$. It should be apparent to the reader that the expectation value of $b_M(X)$ is dependent on t as well as y due to the time dependence of the Y field. Thus the scale factor will exhibit a considerably more intricate behavior than simply its $a(t)$ dependence.

The Fourier expansion of the Y field is:

$$Y^i(z) = \int d^3p\, N_0(\omega) \sum_{\lambda=1}^{2} \varepsilon^i(p, \lambda)[a(p,\lambda)\, e^{-ip\cdot z} + a^\dagger(p,\lambda)\, e^{ip\cdot z}] \tag{9.42}$$

where z^μ will be set equal to y^μ later, and where

$$N_0(\omega) = [(2\pi)^3 2\omega]^{-1/2} \tag{9.43}$$

and

$$\omega = (\mathbf{p}^2)^{½} = p^0 \tag{9.44}$$

with $\vec{\varepsilon}(p, \lambda)$ being the polarization unit vectors for λ = 1, 2 and $\eta_{\mu\nu}p^\mu p^\nu = 0$. The expectation value of Y between the |BB, T> states is:

$$<BB, T|Y^i(z)|BB, T> = \int d^3p\ N_0(\omega)f(\omega,T)[e^{-ip\cdot z} + e^{ip\cdot z}] \sum \varepsilon^i(p, \lambda) \tag{9.45}$$

The evaluation of eq. 9.45 (and spherical symmetry) gives

$$<BB, T|Y^i(z)|BB, T> = \hat{y}^i \int d^3p\ N_0(\omega)f(\omega,T)[e^{-ip\cdot z} + e^{ip\cdot z}] \sum_\lambda \hat{\mathbf{z}}\cdot \varepsilon(p, \lambda) \tag{9.46}$$

$$\equiv \hat{\mathbf{z}}^i\ Y_{BB}(t, z)$$

where $\hat{\mathbf{z}} = \vec{\mathbf{z}}/|\vec{\mathbf{z}}|$ is the unit three-vector in the direction of $\vec{\mathbf{z}}$, z = $|\vec{\mathbf{z}}|$, and $p\cdot z = \omega(t - z\cos\theta)$. We define a spatial coordinate system – choosing the z-axis parallel to $\vec{z}$. Then we have

$$\vec{z} = (0, 0, z) \tag{9.47}$$

$$\vec{p} = (\sin\theta\cos\phi\ ,\ \sin\theta\sin\phi,\ \cos\theta) \tag{9.48}$$

$$\vec{\varepsilon}(p,1) = (\cos\theta\cos\phi\ ,\ \cos\theta\sin\phi,\ -\sin\theta) \tag{9.49}$$

$$\vec{\varepsilon}(p,2) = (-\sin\phi\ ,\ \cos\phi,\ 0) \tag{9.50}$$

with the result (taking account of eq. 9.46)

$$Y_{BB}(t, z) = <BB, T|\ \hat{\mathbf{z}}\cdot \mathbf{Y}(t, z)\ |BB, T>$$

$$= 2\pi \int_0^\infty d\omega\ \omega^2 N_0(\omega)f(\omega,T) \int_0^\pi d\theta\ \sin^2\theta\ [e^{-ip\cdot z} + e^{ip\cdot z}] \tag{9.51}$$

where $p\cdot z = \omega(t - z\cos\theta)$ with z = $|\vec{z}|$. We will develop integral representations and approximations to Y_{BB} in a later section.

9.2.4 Expectation Values of the Scale Factor A(t, X) and the Invariant Interval $d\tau^2$

The scale factor

$$b_M(X) = :2\alpha M_c k^{-1/2}[1 + \alpha^2 M_c^2 X^2]^{-1}: \quad (9.52)$$

is a normal-ordered q-number expression. We can formally expand (define) this expression as a power series of normal-ordered powers of X^2 and then evaluate it between blackbody coherent states. First we note that

$$<BB, T|:Y^{i1}(z)Y^{i2}(z)Y^{i3}(z)Y^{i4}(z) \ldots Y^{in}(z):|BB, T> = \hat{z}^{i1}\hat{z}^{i2}\hat{z}^{i3}\hat{z}^{i4} \ldots \hat{z}^{in}(Y_{BB}(t, z))^n \quad (9.53)$$

We now set $z^\mu = y^\mu$, and use Y_{BB} to represent $Y_{BB}(t, \mathbf{y})$:

$$Y_{BB} \equiv Y_{BB}(t, \vec{y}) \quad (9.54)$$

Thus

$$b_{BB}(y, t) = <BB, T|b_M(X)|BB, T>$$

$$= 2\alpha M_c k^{-1/2}\{1 + \alpha^2[M_c^2 y^2 + 2i\, yY_{BB} - Y_{BB}^2/M_c^2]\}^{-1} \quad (9.55)$$

and

$$B_{BB}(t, y) = <BB, T|B(t, X)|BB, T> = a(t)b_{BB}(y, t) \quad (9.56)$$

The expectation value of the q-number invariant interval (eq. 9.29) is the c-number expression:

$$d\tau_{BB}^2 = <BB, T|d\tau^2|BB, T>$$

$$= dt^2 - B_{BB}^2(t, y)[dX_{BB}^2 + X_{BB}^2(d\theta^2 + \sin^2\theta\, d\varphi^2)] \quad (9.57)$$

where

$$X_{BB} = y + iY_{BB}/M_c^2 \quad (9.58)$$

and

$$dX_{BB} = dy(1 + iM_c^{-2}\partial Y_{BB}/\partial y) \quad (9.59)$$

The appearance of Y_{BB} in the expression for the invariant interval (eq. 9.57) has two effects: it introduces complex space-time into the model and the generalized Robertson-Walker metric is no longer equivalent to the Robertson-Walker metric for all a as it would be in the classical case.

We will see that Y_{BB} approaches zero at large times thus yielding the conventional Robertson-Walker models. But at small times of the order of the Planck time near the Big Bang we enter a brave new world of complex space-time. We will investigate the nature of this new complex world in the succeeding sections.

9.2.5 Representation and Approximations for $Y_{BB}(t, z)$

The angle integral in eq. 9.51 can be performed to yield

$$Y_{BB}(t, z) = \pi^{1/2} z^{-1} \int_0^\infty d\omega\, \omega^{-1}(e^{\omega/\kappa T} - 1)^{-1/2} \cos(\omega t) J_1(\omega z) \tag{9.60}$$

where $J_1(\omega z)$ is a Bessel function using 3.915.5 of Gradshteyn (1965).

9.2.5.1 Some Representations of Y_{BB}

The integral in eq. 9.60 does not appear to be simply expressible in terms of standard transcendental functions. A series representation of the integral can be obtained by expanding the exponential factor due to the Planck distribution:

$$Y_{BB}(t, z) = \tfrac{1}{2}\pi^{1/2}\kappa T \sum_{n=0}^{\infty} (2n)![2^{2n}(n!)^2]^{-1}\{(2n+1+ 2i\kappa Tt)^{-1} F(\tfrac{1}{2}, 1; 2; -[\kappa Tz/(n+\tfrac{1}{2} + i\kappa Tt)]^2) + (2n+1 - 2i\kappa Tt)^{-1} F(\tfrac{1}{2}, 1; 2; -[\kappa Tz/(n+\tfrac{1}{2} - i\kappa Tt)]^2)\} \tag{9.61}$$

where $F(a, b; c; w)$ is a hypergeometric function.[92]

Using an integral representation[93]

$$F(a, b; c; w) = \Gamma(c)[\Gamma(b)\Gamma(c - b)]^{-1} \int_0^1 dt\, t^{b-1}(1 - t)^{c-b-1}(1 - tw)^{-a}$$

for $F(\tfrac{1}{2}, 1; 2; w)$ we see eq. 9.61 can be written in terms of simpler algebraic expressions:

---

[92] Based on the integral 6.621.1 on p. 711 of Gradshteyn (1965).

[93] Magnus (1949) p. 8.

$$Y_{BB}(t, z) = \tfrac{1}{2}\pi^{1/2}(\kappa T z^2)^{-1} \sum_{n=0}^{\infty} (2n)![2^{2n}(n!)^2]^{-1}\{[(n+ \tfrac{1}{2} + i\kappa Tt)^2 + (\kappa Tz)^2]^{1/2} + \\ + [(n+ \tfrac{1}{2} - i\kappa Tt)^2 + (\kappa Tz)^2]^{1/2} - (2n+1)\} \qquad (9.62)$$

Eq. 9.62 shows the limit of $Y_{BB}(t, z)$ for large t is

$$Y_{BB}(t, z) \to \pi^{1/2}\kappa T \sum_{n=0}^{\infty} (2n)![2^{2n+2}(n!)^2]^{-1}(2n + 1)[(n + \tfrac{1}{2})^2 + (\kappa Tt)^2]^{-1} \to 0 \qquad (9.63)$$

if $tT \to \infty$ as $t \to \infty$ as we see in cosmological models (see section 9.4).

Thus

$$(b_{BB}(y, t))^2 dX^2 \to 2\alpha M_c^2 k^{-1}[1 + \alpha^2 M_c^2 y^2]^{-2} dy^2 \equiv 2\alpha k^{-1}[1 + \alpha^2 \check{r}^2]^{-2} d\check{r}^2 \qquad (9.64)$$

for large t showing the Two-Tier cosmological model becomes a Robertson-Walker model at large times. However, the Two-Tier standard cosmological model is very different at small times of the order of the Planck time near the Big Bang point.

9.2.5.2 Approximate Solution for Y_{BB}

The integral representation and power series representation of $Y_{BB}(t, y)$ do not reveal the physical behavior of the model for small times t and distances y. Therefore we will examine an approximation for $Y_{BB}(t, y)$ for ranges of y, t and T that are relevant for our considerations. We begin by scaling the integration variable in eq. 9.60 with the result:

$$Y_{BB}(t, y) = \pi^{1/2} y^{-1} \int_0^{\infty} d\omega\, \omega^{-1}(e^{\omega} - 1)^{-1/2}\cos(\omega\kappa Tt) J_1(\omega\kappa Ty) \qquad (9.65)$$

The blackbody exponential factor $(e^{\omega} - 1)^{-1/2}$ in the integrand of $Y_{BB}(t, y)$ enables the leading order approximate behavior of $Y_{BB}(t, y)$ to be determined for $0 \le t \lessapprox 10^{108}$ s – for all time, practically speaking. In a later section (section 9.3.4) we will see that our approximation to the integral in eq. 9.65 is consistent with the solution that we obtain for Y_{BB}, for the scale factor and thus for the temperature T. The approximations that we will make in eq. 9.65 are

$$\cos(\omega\kappa \mathrm{T}t) \approx 1 \tag{9.66}$$

$$J_1(\omega\kappa \mathrm{T}y) \approx \omega\kappa \mathrm{T}y/2 \tag{9.67}$$

They are based on κTt $\ll 1$ and κTy $\ll 1$ for all y ($0 \le y \le M_c^{-1}$). The exponential factor tends to limit contributions to the integral to small ω. After making these approximations we find

$$Y_{BB}(t, y) \simeq \tfrac{1}{2}\, \pi^{1/2}\kappa \mathrm{T} \int_0^\infty d\omega\, (e^\omega - 1)^{-1/2}$$

$$\simeq \pi^{3/2}\kappa \mathrm{T}/2 \tag{9.68}$$

The limit as t gets large can also be approximately determined from eq. 9.65. For large t such that κTy is small (and ω is small due to the exponential Planck distribution factor) we can again approximate the Bessel function with its leading power series expansion term and the exponential factor can again be approximated by $e^\omega - 1 \approx \omega$ so that eq. 9.65 becomes approximately

<u>As t → ∞:</u>

$$Y_{BB}(t, y) \simeq \pi^{1/2} 2^{-1}\, \kappa \mathrm{T} \int_0^\infty d\omega\, \omega^{-1/2} \cos(\omega\kappa \mathrm{T}t)$$

$$= \pi 2^{-3/2} [\kappa \mathrm{T}/t]^{1/2} \tag{9.69}$$

using 3.751.2 of Gradshteyn (1965). We note that, while κTy is small, κTt could possibly have been large in either a matter-dominated or radiation-dominated universe since it grows as t to a positive power (see section 9.3.) *However, since $Y_{BB}(t, y)$ approaches zero for large times its impact can only be seen in the initial formative stages of the universe near t = 0.*

9.3 The Scale Factor a(t) Near t = 0

The "time factor" a(t) of the scale factor $B_{BB}(t, y)$ appears in

$$B_{BB}(t, y) = \langle BB, T|B(t, X)|BB, T\rangle = a(t) b_{BB}(y, t) \tag{9.70}$$

and is determined by the classical Einstein equation:

$$\dot{a}^2 - 8\pi G\rho a^2/3 = -k \tag{9.71}$$

As we have argued earlier, the source determining a(t) for small times in the neighborhood of t = 0 (the time of the Big Bang) is a large, macroscopic, classical density $\rho(t)$ and thus a(t) may be considered to be a c-number quantity determined by the c-number Einstein equation to good approximation. This approximation should continue to hold even if this macroscopic density becomes enormous as t → 0. The quantum effects near t = 0 in the Two-Tier model, that we have developed, appear in the factor $b_{BB}(y, t)$ that we evaluated in previous sections.

9.4 A Complex Blackbody Temperature Near t = 0

The blackbody temperature T for relativistic particles (presumably the dominant type of particles near t = 0) is inversely proportional to the scale factor. At large times the blackbody temperature has the form

$$T = T_0/a(t) \tag{9.72}$$

where T_0 is a constant.

At times in the neighborhood of t = 0 (the Big Bang) space has three complex dimensions in the Two-Tier model. Temperature can be viewed as a measure of the root mean square speed (or the "average energy") of the components of the perfect fluid that we have assumed. In the case of a gas of particles of average energy E:

$$T = E/(3k/2) \tag{9.73}$$

In a complex space it is quite natural for the root mean squared speed to be complex as well. As a result complex temperatures naturally follow. Thus we will define

$$T = T_0/B_{BB}(t, y) \tag{9.74}$$

for all time since $t = 0$. Since $B_{BB}(t, y)$ approaches $M_c a(t) b(\check{r})$ at large times we find its large time behavior is consistent with those of standard Robertson-Walker models. At times near $t = 0$, the blackbody temperature T is complex since space is complex and complex kinetic energy is allowed.

The appearance of complex quantities in the preceding paragraph is remedied by the use of the Reality group which maps complex quantities to their physical real values. In the case of the complex temperature T above, the corresponding physical temperature is its absolute value (obtained by multiplying T by a phase factor from the Reality group U(4))

$$T_{physical} = |T_0/B_{BB}(t, y)| \qquad (9.74a)$$

We shall use eq. 9.74 for the temperature transforming the results below, afterwards, to real values using the 4-dimensional Reality group.

9.5 The Nature of the Universe Near t = 0

At this point we are ready to examine the Two-Tier model for the Big Bang period.

9.5.1 Behavior of the Complete Scale Factor B(t, y) Near t = 0

The behavior of the expectation value of the scale factor B(t, y), under the assumption that the Y quanta have a blackbody spectrum, is described by the equations:

$$b_{BB}(y, t) = 2\alpha M_c k^{-1/2}\{1 + \alpha^2[M_c^2 y^2 + 2i\, yY_{BB} - Y_{BB}^2/M_c^2]\}^{-1} \qquad (9.75)$$

$$B_{BB}(t, y) = \langle BB, T|B(t, X)|BB, T\rangle = a(t)b_{BB}(y, t) \qquad (9.76)$$

$$Y_{BB}(t, y) \simeq \pi^{3/2}\kappa T/2 \qquad (9.77)$$

$$a(t) = [2\pi G\rho_0 n^2/3]^{1/n} t^{2/n} \qquad (9.78)$$

$$T = T_0/B_{BB}(t, y) \qquad (9.79)$$

as t → 0. We will set $\alpha = 1$ in the interests of simplicity knowing that this value results in a metric fully equivalent to the Robertson-Walker metric at large times. (Other values of α[94] would also result in a metric equivalent to the Robertson-Walker metric at large times after a re-scaling of the radial coordinate.) Thus we may write

$$B_{BB}(t, y) \cong 2k^{-1/2}M_c a(t)\{1 + M_c^2 y^2 + iy\pi^{3/2}\kappa T_0/B_{BB} - \pi^3\kappa^2 T_0^2/(4M_c^2 B_{BB}^2)\}^{-1} \qquad (9.80)$$

---

[94] It is **not** the fine structure constant.

This quadratic algebraic equation for B_{BB} has the solutions:

$$B_{BB}(t, y) \cong (1 + M_c^2y^2)^{-1}\{-i\varkappa M_cy + k^{-1/2}M_ca(t) \pm [\varkappa^2 - 2i\varkappa yk^{-1/2}M_c^2a(t) + k^{-1}M_c^2a^2(t)]^{1/2}\} \tag{9.81}$$

with

$$\varkappa = \pi^{3/2}\kappa T_0/(2M_c) \tag{9.82}$$

As t gets very large we obtain the equivalent of the Robertson-Walker metric scale factor in this approximation if we choose the plus sign in eq. 9.81 (assuming $a^2(t)$ becomes very large so other terms within the square root can be neglected):

$$B_{BB}(t, y) \rightarrow 2k^{-1/2}M_ca(t)/(1 + M_c^2y^2) \tag{9.83}$$

Thus we must choose the plus sign in eq. 9.81:

$$B_{BB}(t, y) \cong (1 + M_c^2y^2)^{-1}\{-i\varkappa M_cy + k^{-1/2}M_ca(t) + [\varkappa^2 - 2i\varkappa yk^{-1/2}M_c^2a(t) + k^{-1}M_c^2a^2(t)]^{1/2}\} \tag{9.84}$$

At t = 0 (the Big Bang) eq. 9.84 simplifies to (assuming a(0) = 0)

$$B_{BB}(0, y) \cong [\varkappa - i\varkappa M_cy]/(1 + M_c^2y^2) \tag{9.85}$$

For small y, the real part of $B_{BB}(0, y)$ is a constant and the imaginary part of $B_{BB}(0, y)$ is proportional to y.

9.5.2 The Expectation Value of the Scale Factor $A_{BB}(0, y)$ near t = 0

Eq. 9.85 gives the approximate behavior of the expectation value of the scale factor $B_{BB}(0, y)$ near t = 0 as a function of y. If we compare eq. 9.85 with the mechanism described in eqns. 9.1.6 – 9.1.9 of section 9.1 for cancelling the a(t) factor within the complete scale factor we see that we have found the blackbody spectrum of the Y quanta implements this mechanism. The solution can be written in the form:

$$A_{BB}(t, y) = M_c^{-1}B_{BB}(t, y) \cong \beta_0(y) + \beta_1(y)a(t) + \ldots \tag{9.86}$$

$$\beta_0(y) = \varkappa(1 - iM_cy)/[M_c(1 + M_c^2y^2)] \tag{9.87}$$

$$\beta_1(y) = k^{-½}(1 - iM_c y)/(1 + M_c^2 y^2) \quad (9.88)$$

Eqns. 9.86–9.88 are expressed in terms of the y variable. They can be expressed in terms of ř as:

$$A(t, ř) = a(t)b(ř) = 2\alpha k^{-½}a(t)[1 + \alpha^2 ř^2]^{-1} \quad (9.89)$$

with $\alpha = 1$. Evidently, we have $M_c y \equiv ř$ at the level of approximation that we are using. Furthermore we can use

$$ř = \{[1 - (1 - kr^2)^{½}]/[1 + (1 - kr^2)^{½}]\}^{½} \quad (9.90)$$

to express ř in terms of the Roberson-Walker radial coordinate r (from appendix 9-A.) Thus we find that the Robertson-Walker scale factor a(t) becomes

$$a(t) \rightarrow (1 + M_c^2 y^2)(2k^{-½})^{-1}A_{BB}(t, y) \equiv a_{BBRW}(t, ř) \quad (9.91)$$
$$a_{BBRW}(t, ř) \cong \beta_{0RW}(ř) + \beta_{1RW}(ř)a(t) + \ldots \quad (9.92)$$

using the subscript "RW" to denote quantities scaled to the standard Robertson-Walker metric, with

$$\beta_{0RW}(ř) = \varkappa(1 - iř)/[2k^{-½}M_c] \quad (9.93)$$
$$\beta_{1RW}(ř) = (1 - iř)/2 \quad (9.94)$$

where ř is specified by eq. 9.90.

Using the Reality group we can "rotate" the complex scale factor which is a factor in coordinate expressions to a real value—its absolute value

$$a(t) \rightarrow (1 + M_c^2 y^2)(2k^{-½})^{-1}|A_{BB}(t, y)| \equiv |a_{BBRW}(t, ř)| \quad (9.91a)$$

We will study the implications of this scale factor in the following chapters.

Appendix 9-A. Derivation of the Extended Robertson-Walker Model

This appendix provides the derivation of a generalization of the Roberson-Walker solution of General Relativity that we used in chapter 9 in our discussion of the Quantum Big Bang Theory originally presented in Blaha (2004). This appendix is extracted from chapter 8 of Blaha (2004) as necessary background information for chapter 9 and the following chapters.

The generalization derived here has not been derived before, to our knowledge, because at the level of c-number General Relativity it is fully equivalent to the known Robertson-Walker model. However when the theory has q-number parts introduced in a physically meaningful way, as we do, it is no longer equivalent to the c-number Robertson Walker model.

9-A.1 The Robertson-Walker Metric

Much of the current modeling of the evolution and properties of the universe is based on the assumption of a Robertson-Walker metric which is used in the Einstein equations to obtain a first order differential equation for the scale factor $R(t)$:

$$\dot{R}^2 + k = 8\pi G\rho R^2/3 \qquad (9\text{-A.1.1})$$

where k is a factor in the three-dimensional spatial curvature of the Robertson-Walker metric:

$$K_3(t) = k/R^2(t) \qquad (9\text{-A.1.2})$$

Eq. 9-A.1.2 suggests that accurate measurements of the Hubble constant and other cosmological quantities could lead to an accurate determination of the curvature, the time dependence of the Hubble constant, and of R(t).

The form of the Robertson-Walker metric (We consider a real space-time only in this chapter.) follows from the assumption of a maximally symmetric three-dimensional subspace whose metric has eigenvalues of the same sign (negative in our formalism) residing within a four-dimensional space-time with one positive eigenvalue and three negative eigenvalues. A maximally symmetric space is isotropic and homogeneous.[95]

Although the Robertson-Walker metric does not appear to embody the concept of an absolute space-time the general arguments presented in chapter 2 of Blaha (2004) indicate that it does in fact implicitly define an absolute reference frame.[96]

Therefore it is sensible to inquire whether a more general metric – a generalization of the Robertson-Walker metric – might be worth investigating – particularly in view of the major unexplained mysteries of Dark Energy and Dark Matter as well as other new data showing the existence of massive black holes and quite mature galaxies shortly after (two or three billion years) the origin of the universe. The pile-up of mysteries from WMAP, SDSS and other sources indicates a reconsideration of fundamental assumptions may be worthwhile.

Therefore we will examine the "simplest" generalization of the Robertson-Walker metric in this chapter with a view towards elucidating some of these mysteries. More importantly, we will use this generalization in chapter 9 and subsequent chapters to develop a non-singular, Two-Tier formulation of the dynamics of the universe "at the beginning of time" – the Big Bang – taking account of quantum effects.

From the point of view of the definition of maximally symmetric subspaces the most immediate generalization of the Robertson-Walker metric is to assume a maximally symmetric *two-dimensional* subspace within a four-dimensional space-time. The general form of the metric in this case is:

$$d\tau^2 = A_{tt}(r,t)\, dt^2 + 2A_{rt}(r,t)\, dt\, dr + A_{rr}(r,t)\, dr^2 + B(r,t)(d\theta^2 + \sin^2\theta d\varphi^2) \qquad \text{(9-A.1.3)}$$

where A_{ik} is a 2×2 symmetric matrix with one positive and one negative eigenvalue, and $B(r,t)$ is a negative function of r and t.

---

[95] See Chapter 13 of Weinberg (1972) for a detailed discussion.

[9696] This conclusion is now obvious due to the existence of the Flatverse. We note there can only be one absolute reference frame up to a Lorentz transformation since the sets of inertial reference frames of two absolute reference frames must be the same. This implies any two absolute reference frames must be related by a Lorentz transformation since absolute frames are necessarily flat inertial frames.

9-A.2 A Generalization of the Roberson-Walker Metric

We shall consider a generalization of the Robertson-Walker metric (eq. 5.5.1 of Blaha (2004)), which is a special case of eq. 9-A.1.3 that preserves the overall form of the Robertson-Walker metric but allows the scale factor a(t) to depend on r as well as t:

$$R(t) \rightarrow A_0(t, r) \qquad (9\text{-A.2.1})$$

The generalized metric that we will analyze is embodied in the invariant interval expression:

$$d\tau^2 = dt^2 - A_0^2(t, r)[dr^2/(1 - kr^2) + r^2(d\theta^2 + \sin^2\theta\, d\varphi^2)] \qquad (9\text{-A.2.2})$$

The introduction of a dependence on the radius r in $A_0(t, r)$ in eq. 9-A.2.2 eliminates the homogeneity of the three-dimensional spatial subspace reducing it to a two-dimensional maximally symmetric subspace.

We note that the general solution of the Einstein equations for the standard case of a perfect fluid lead to the usual view of the expansion of the universe (after the Big Bang Epoch), Hubble's law, the red shifts of radiation from distant sources of radiation, and the Cosmic Microwave Background (CMB) radiation.

9-A.3 The Einstein Equations for the Generalized Robertson-Walker Metric

The Einstein equations can be written

$$R_{\mu\nu} = -8\pi G S_{\mu\nu} \qquad (9\text{-A.3.1a})$$

$$S_{\mu\nu} = T_{\mu\nu} - \tfrac{1}{2} g_{\mu\nu} T^{\sigma}{}_{\sigma} \qquad (9\text{-A.3.1b})$$

where $R_{\mu\nu}$ is the Ricci tensor and $T_{\mu\nu}$ is the energy-momentum tensor. Assuming the energy-momentum tensor has the form of the energy-momentum tensor of a perfect fluid with the only non-zero components:

$$T_{tt} = \rho\, g_{tt} \qquad (9\text{-A.3.2})$$

$$T_{rr} = -\, p g_{rr} \qquad (9\text{-A.3.3})$$

$$T_{\theta\theta} = -\, p g_{\theta\theta} \qquad (9\text{-A.3.4})$$
$$T_{\phi\phi} = -\, p g_{\phi\phi} \qquad (9\text{-A.3.5})$$

where ρ is the density and p is the pressure, then the non-zero components of $S_{\mu\nu}$ are:

$$S_{tt} = ½\,(\rho + 3p)\, g_{tt} \qquad (9\text{-A.3.6})$$
$$S_{rr} = -½\,(\rho - p) g_{rr} \qquad (9\text{-A.3.7})$$
$$S_{\theta\theta} = -½\,(\rho - p) g_{\theta\theta} \qquad (9\text{-A.3.8})$$
$$S_{\phi\phi} = -½\,(\rho - p) g_{\phi\phi} \qquad (9\text{-A.3.9})$$

In particular, the fact that

$$S_{tr} = 0 \qquad (9\text{-A.3.10})$$

for a perfect fluid results in an important simplification in the solution of the Einstein equations for this case.

The density $\rho = \rho(t)$ and the pressure $p = p(t)$ are assumed to be solely functions of time t as is usual in the case of a perfect fluid.

9-A.4 The Differential Equations for the Generalized Scale Factor A(t, r)

The dependence of the scale factor A_0 on both r and t leads to a significantly more complicated calculation of the Ricci tensor. We start by noting

$$g_{tt} = 1 \qquad g_{rr} = -A_0^2/(1 - kr^2) \qquad g_{\theta\theta} = -A_0^2 r^2 \qquad g_{\phi\phi} = -A_0^2 r^2 \sin^2\theta \qquad (9\text{-A.4.1})$$

Despite the dependence of A_0 on both t and r in the generalized case a direct calculation of the tt Ricci tensor component yields the familiar expression:

$$R_{tt} = g_{tt}\, 3\ddot{A}_0/A_0 \qquad (9\text{-A.4.2})$$

where we use dots over A_0 to indicate partial derivatives with respect to time:

$$\ddot{A}_0 \equiv \partial^2 A_0/\partial t^2 \qquad (9\text{-A.4.3})$$

However, the tr-component of the Ricci tensor R_{tr}, which is zero in the case of the ordinary Robertson-Walker metric, is non-zero in the more general case under consideration:

$$R_{tr} = 2\,\partial(\partial A_0/\partial t)/\partial r \tag{9-A.4.4}$$

The corresponding Einstein equation is

$$R_{tr} = 2\,\partial(A_0^{-1}\,\partial A_0/\partial t)/\partial r = -8\pi G S_{tr} = 0 \tag{9-A.4.5}$$

for a perfect fluid. Eq. 9-A.4.5 implies that $A_0(t, r)$ factorizes:

$$A_0(t, r) = a(t)b_0(r) \tag{9-A.4.6}$$

This factorization results in a substantial simplification in the non-linear Einstein equations considered next, which are shown to be separable in the radial and time variables.

Before proceeding to the consideration of the remaining Einstein equations it is convenient to redefine the radial coordinate using

$$\check{r}\,b(\check{r}) = r\,b_0(r) \tag{9-A.4.7}$$

and

$$d\check{r}/dr = b_0(r)[b(\check{r})(1 - kr^2)^{½}]^{-1} = \check{r}\,[r(1 - kr^2)^{½}]^{-1} \tag{9-A.4.8}$$

While this change of coordinates does not change the physical content of the theory it does lead to simpler Einstein equations. The change of radial coordinate results in a new form of the invariant interval:

$$d\tau^2 = dt^2 - a^2(t)b^2(\check{r})[d\check{r}^2 + \check{r}^2(d\theta^2 + \sin^2\theta\,d\varphi^2)] \tag{9-A.4.9}$$

The new radial coordinate $\check{r}$ is related to the old radial coordinate by

$$\check{r} = \{[1 - (1 - kr^2)^{½}]/[1 + (1 - kr^2)^{½}]\}^{½} \tag{9-A.4.10}$$

and

$$r = 2k^{-½}\check{r}(1 + \check{r}^2)^{-1} \tag{9-A.4.11}$$

Note the range of ř is [0, 1].

A direct calculation of the Ricci tensor for the metric in eq. 9-A.4.9 with A(t, ř) defined as:

$$A \equiv A(t, ř) = a(t)b(ř) \qquad (9\text{-A.4.12})$$

leads to the following Einstein equations (remembering our flat space cartesian metric is $\eta_{tt} = +1$ and $\eta_{ij} = -\delta_{ij}$ for i, j = spatial indices):

$$R_{tt} = g_{tt}3\ddot{A}/A = -8\pi G S_{tt} = -4\pi G(\rho + 3p) \qquad (9\text{-A.4.13})$$

$$R_{tř} = 2\,\partial(A^{-1}\,\partial A/\partial t)/\partial ř = -8\pi G\, S_{tř} = 0 \qquad (9\text{-A.4.14})$$

$$R_{řř} = g_{řř}[\ddot{A}/A + 2(\dot{A}/A)^2 - 2(řA^2)^{-1}\,\partial(ř\,A'/A)/\partial ř]$$

$$= -8\pi G\, S_{řř} = 4\pi G\, g_{řř}(\rho - p) \qquad (9\text{-A.4.15})$$

$$R_{\theta\theta} = g_{\theta\theta}[\ddot{A}/A + 2(\dot{A}/A)^2 - A''/A^3 - 3A'/(řA^3)^{-1}]$$

$$= -8\pi G\, S_{\theta\theta} = 4\pi G\, g_{\theta\theta}(\rho - p) \qquad (9\text{-A.4.16})$$

$$R_{\phi\phi} = g_{\phi\phi}[\ddot{A}/A + 2(\dot{A}/A)^2 - A''/A^3 - 3A'/(řA^3)^{-1}]$$

$$= -8\pi G\, S_{\phi\phi} = 4\pi G\, g_{\phi\phi}\,(\rho - p) \qquad (9\text{-A.4.17})$$

where

$$A' = \partial A/\partial ř \qquad (9\text{-A.4.18})$$

and

$$A'' = \partial^2 A/\partial ř^2 \qquad (9\text{-A.4.19})$$

9-A.5 The Solution for the Generalized Scale Factor A(t, r)

We begin by substituting eq. 9-A.4.13 in eq. 9-A.4.15, and then substituting the factorization of A (eq. 9-A.4.12). The result is a separable equation:

$$\dot{a}^2 - 8\pi G\rho a^2/3 - (\check{r}b^2)^{-1}\,\partial(\check{r}\,b'/b)/\partial\check{r} = 0 \qquad (9\text{-A.5.1})$$

Since the first two terms in eq. 9-A.5.1 are solely functions of t while the third term is solely a function of ř we obtain the separated equations:

$$\dot{a}^2 - 8\pi G\rho a^2/3 = -k \qquad (9\text{-A.5.2})$$

and

$$k + (\check{r}b^2)^{-1}\,\partial(\check{r}\,b'/b)/\partial\check{r} = 0 \qquad (9\text{-A.5.3})$$

where k is a separation constant which we provisionally identify with the curvature parameter k in eq. 9-A.2.2.

Eq. 9-A.5.2 is precisely the equation used for the time dependent scale factor in current cosmological models (eq. 5A.3.2) using the Robertson-Walker metric. Therefore its solution, which depends on the time dependence of the energy density, will be the same as that of the corresponding conventional cosmological model for the same density.

Eq. 9-A.5.3 is a differential equation for the *spatial* expansion scale factor b(ř). Under the assumption of a perfect fluid it depends solely on the constant k and is independent of the details of the perfect fluid (i.e. its density and pressure). There are two solutions of the second order non-linear differential equation for b(ř). These solutions can be written as

$$b_1(\check{r}) = 2\gamma\delta k^{-1/2}[\delta^2 + \gamma^2\,\check{r}^2]^{-1} \qquad (9\text{-A.5.4})$$

where γ and δ are constants, and

$$b_2(\check{r}) = \sigma k^{-1/2}[\check{r}\,(\varsigma \pm i\sigma \ln \check{r})]^{-1} \qquad (9\text{-A.5.5})$$

where σ and ς are constants.

b_1(ř) (eq. 9-A.5.4) is the only physically acceptable solution for the case of a perfect fluid with energy-momentum tensor specified by eqns. 9-A.3.2 – 9-A.3.5. Reason: The solution b_1(ř) satisfies eqns. 9-A.4.16 and 9-A.4.17 while the other solution b_2(ř) does not satisfy these equations. A necessary and sufficient condition for b(ř) to satisfy eqns. 9-A.4.16 and 9-A.4.17 is that

$$b'' - \check{r}^{-1} b' - 2\, b'^2/b = 0 \qquad (9\text{-A.5.6})$$

where ′ denotes a derivative with respect to ř. Eq. 9-A.5.6 follows from subtracting the coefficients of the metric tensor component factors in eq. 9-A.4.15 from the coefficients of the metric tensor component factors in eq. 9-A.4.16 (or eq. 9-A.4.17). The solution $b_1(\check{r})$ satisfies eq. 9-A.5.6 for all values of γ and δ. The solution $b_2(\check{r})$ does not satisfy eq. 9-A.5.6 for any choice of α and β except the trivial choice $\alpha = 0$ and is therefore physically irrelevant. It might have some relevance in the case of a non-perfect fluid. We will not investigate that possibility in the present work.

9-A.6 The Solution Expressed in the Original Radial Coordinate

The solution that we have obtained for the generalized Roberson-Walker case with radial coordinate ř can be related back to the generalized Robertson-Walker solution using the original radial coordinate r. Eq. 9-A.4.7 implies

$$b_0(r) = \check{r}\, b(\check{r})/r \qquad (9\text{-A.6.1})$$

$$= 2\alpha\, [1 + \alpha^2 + (1 - \alpha^2)(1 - kr^2)^{1/2}]^{-1} \qquad (9\text{-A.6.2})$$

using eqns. 9-A.4.10 and 9-A.5.4 and

$$\alpha = \gamma/\delta \qquad (9\text{-A.6.3})$$

An important special case of eq. 9-A.6.2 is the case where $\alpha = 1$. In this case we find eq. 9-A.6.2 becomes

$$b_0(r) = 1 \qquad (9\text{-A.6.4a})$$

and

$$A_0(t, r) = a(t) \qquad (9\text{-A.6.4b})$$

thus *recovering the normal Robertson-Walker solution exactly as a special case from eqns. 9-A.2.2, 9-A.4.6, and 9-A.5.2.* In this case we see

$$b(\check{r}) = 2k^{-1/2}(1 + \check{r}^2)^{-1} \qquad (9\text{-A.6.5})$$

from eq. 9-A.5.4 in the ř, θ, ϕ coordinate system. We considered this case within the expanded framework of a quantized model of the beginning of the universe in chapter 9 where it has non-trivial consequences.

9-A.7 Equivalence of the General Solution with the Original Robertson-Walker Solution

The general solution of the Einstein equations in the case of a perfect fluid (eqns. 9-A.4.13 – 9-A.4.17) for the scale factor A(t, ř) in the generalized metric specified by eq. 9-A.4.9

$$d\tau^2 = dt^2 - A^2(t, \check{r})[d\check{r}^2 + \check{r}^2(d\theta^2 + \sin^2\theta\, d\varphi^2)] \qquad (9\text{-A.4.9})$$

is given by

$$A(t, \check{r}) = a(t)b(\check{r}) = 2\alpha k^{-1/2}a(t)[1 + \alpha^2\check{r}^2]^{-1} \qquad (9\text{-A.7.1})$$

where a(t) is the solution of the standard equation (eq. 9-A.5.2) for the time dependent scale factor in the case of the Robertson-Walker metric. We now note that if we define a new radial vector

$$r = \alpha\check{r} \qquad (9\text{-A.7.2})$$

then

$$d\tau^2 = dt^2 - a^2(t)4k^{-1}[1 + r^2]^{-1}\,[dr^2 + r^2(d\theta^2 + \sin^2\theta\, d\varphi^2)] \qquad (9\text{-A.7.3})$$

Comparing this invariant interval expression with the α = 1 expression for b(ř) given in eq. 9-A.6.5 we conclude that we have proved the following theorem:

Theorem: The solution of the Einstein equations for the case of a perfect fluid for the generalized Robertson-Walker metric (eq. 9-A.2.2) is equivalent to a solution of the Einstein equations for the case of a perfect fluid in the case of the Robertson-Walker metric with a scale factor a(t) that is solely dependent on time.

In the case of *classical* gravitation theory the solutions are fully equivalent and related by a simple change of radial coordinates. Thus the homogeneity condition that we had relinquished at the beginning of our discussion is reinstated and the solution of the generalized case is consistent with a *maximally symmetric three-dimensional space*. The origin of the coordinate system can be chosen to be any point in space.

In the case of quantized versions of the Robertson-Walker model and our generalization of it we will see that the solutions are *generally not equivalent*. We explored a particular example of a quantized gravitational model that illustrates this point in chapter 9. The quantized model, which we defined, should be viewed as a first attempt to explore the quantum regime existing at the beginning of time – the Big Bang. The reasonableness of its results suggests that we are on the right track for understanding the Big Bang Epoch.

9-A.8 Hubble's Law in the Generalized Robertson-Walker Model

Our generalized Robertson-Walker metric assumes an inhomogeneous space with some fixed center at ř = 0. Presumably this center was the point at which the Big Bang took place at the beginning of the universe. In this section see how Hubble's Law emerges in the generalized model.

Hubble's law is one of the cornerstones of modern cosmology. While one might think a scale factor that depended on the radius coordinate might not be consistent with Hubble's Law it is easy to show that Hubble's Law is satisfied provided that the scale factor factorizes as required by the tr-component of Einstein's equations (eq. 9-A.4.14) for our generalized Robertson-Walker model.

First we give a simple derivation of Hubble's law for the case of a separable scale factor:

$$A(t, \check{r}) = a(t)b(\check{r}) \qquad (9\text{-A.4.12})$$

under the assumption that some remote galaxy lies on the same radial line as the line from the origin of the space coordinates to our galaxy. The proper distance between the remote galaxy and our galaxy has the form

$$D(t) = D_0 A(t, \check{r}) \qquad (9\text{-A.8.1})$$

The rate of recession of the remote galaxy is then

$$v = dD/dt = D_0 b(\check{r}) da(t)/dt = HD(t) \qquad (9\text{-A.8.2})$$

with

$$H = d \ln a(t)/dt \qquad (9\text{-A.8.3})$$

H is Hubble's constant. If the speed of recession is small then it determines the first-order Doppler shift. If we denote the wavelength of the received radiation as λ_r and the wavelength of the radiation at the source as λ_s then the shift z is

$$z = \lambda_r/\lambda_s - 1 = v/c = HD/c \qquad (9\text{-A.8.4})$$

Eq. 9-A.8.4 is Hubble's Law: the red shift of a galaxy is proportional to its distance. Note Hubble's Law in the generalized case follows from the factorization of A(t, ř) with a(t) satisfying the same differential equation as the Robertson-Walker scale factor.

Since a change of radial coordinate reduces the classical generalized model to the Robertson-Walker model Hubble's Law can be proven in the general case of the non-collinearity of the coordinate origin, source and reception points.

10. Dark Energy, Inflatons and Inflation: The Scale Factor from t = 0 to the Present

10.1 Introduction

This chapter develops a numerical model for the scale factor from the Big Bang to the present. Although the universe that we live in is almost flat according to recent WMAP experimental data, the small curvature of space closes the universe and has significant effects. Therefore we will not use a flat space approximation. Our goal is to obtain an order of magnitude understanding of the evolution of the universe from the beginning. Our calculated numerical quantities appear generally to be of the right order of magnitude. And the physical ideas appear to be consistent with a reasonable view of reality.

Chapter 11 describes the physical implications of our blackbody Y quanta Dark Energy stabilization and expansion mechanisms for the universe. An especially important result of chapter 7 of Blaha (2004) for the expansion of the universe is: **Gravity is a repulsive force (anti-gravity!) at distances less than 9.08 $\times$ 10^{-34} cm.** (See Fig. 7.3.9.3.) Thus the expansion of the universe gets an additional boost from gravity at ultra-short distances.

The data that we use in this chapter and throughout are the combined results of the WMAP and SDSS data[97] based on the assumption of a non-flat space. In particular we use the following values:[98]

$$\begin{aligned}
h &= \text{Hubble parameter} = 0.660 \\
\rho_{cr} &= \text{Critical density} = 1.88h^2 \times 10^{-29}\ \text{g/cm}^3 \\
\Omega_\Lambda &= \text{Dark Energy density}/\rho_{cr} = \rho_{de}/\rho_{cr} = 0.695 \\
\Omega_d &= \text{Dark matter density}/\rho_{cr} = \rho_d/\rho_{cr} = 0.115 \qquad (10.1.1) \\
\Omega_b &= \text{Baryon density}/\rho_{cr} = \rho_b/\rho_{cr} = 0.0230 \\
\Omega_{tot} &= \Omega_m + \Omega_\Lambda = 1.012 \\
\Omega_m &= \text{Matter density}/\rho_{cr} = \rho_m/\rho_{cr} = 0.317
\end{aligned}$$

---

[97] M. Tegmark et al, Phys Rev. **D69**, 103501 (2004): Table IV, column 6. Certain parameters have changed in the past eight years of experiments and observations. However the changes do not significantly change the discussions and conclusions of this chapter and the following chapters.

[98] Physical constants are listed in appendix C.

$$t_0 = t_{now} = \text{Age of universe} = 14.1 \text{ Gyr}$$

The reader is directed to the original paper for other parameters, error bars and a detailed analysis of the data. *We use units where $\hbar = c = 1$ unless stated otherwise.*

In addition to the above input values we will use:[99]

$$\Omega_\gamma = \text{radiation density}/\rho_{cr} = \rho_\gamma/\rho_{cr} = 2.47h^{-2} \times 10^{-5} \quad (10.1.2)$$

and, also, based on an analysis of WMAP[100] data

$$r_{universe}(t_{now}) = \text{current radius of the universe} > 7.4 \times 10^{28} \text{ cm} \quad (10.1.3)$$

10.2 The Behavior of the Scale Factor a(t) after the Big Bang Period

The universe, as we know it today, contains a variety of forms of energy. The current densities of these forms of energy are listed in section 10.1. From them we can develop the form of the total energy density and project it back to the instants after the Big Bang. The Big Bang period is significantly different as we have seen in the preceding chapter.

The total energy density as a function of time is

$$\rho_{tot} = [\Omega_\gamma/a^4(t) + \Omega_m/a^3(t) + \Omega_\Lambda]\rho_{cr} \quad (10.2.1)$$

based on well-known arguments.[101] The time dependence of the scale parameter is given by the Einstein equation:

$$\dot{a}^2 - 8\pi G\rho_{tot}a^2/3 = -k \quad (9\text{-A.5.2})$$

where a(t) is the Robertson-Walker scale factor with $a(t_{now}) = 1$.

[99] Dodelson(2003) p. 41.
[100] N. J. Cornish, D. N. Spergel, G. D. Starkman, and E. Komatsu, Phys. Rev. Lett. **92**, 201302-1 (2004).
[101] Weinberg(1972), Dodelson(2003).

Before proceeding to the solution of eq. 9-A.5.2 we need to obtain a reasonable estimate of the curvature constant k. We can use the Robertson-Walker expression for the radius of the universe

$$r_{universe}(t) = a(t)/k^{½} \qquad (10.2.2)$$

evaluated for the present time where eq. 10.1.3 sets a lower bound on the radius to obtain

$$k^{-½} > 7.4 \times 10^{28} \text{ cm} \qquad (10.2.3)$$

If we *assume* the actual radius of the universe is twice the lower bound, 1.48×10^{29} cm, we get

$$k = (1.48 \times 10^{29})^{-2} \text{ cm}^{-2} = 4.57 \times 10^{-59} \text{ cm}^{-2} \qquad (10.2.4)$$

We will use this value of k in the following sections.

10.2.1 General Form of a(t) Scale Factor Einstein Equation

We find the general form of the scale factor differential equation by combining eqns. 9-A.5.2 and 10.2.1

$$\dot{a}^2 - H_0{}^2 a^2(t)[\Omega_\gamma/a^4(t) + \Omega_m/a^3(t) + \Omega_\Lambda] = -k \qquad (10.2.1.1)$$

where Hubble's constant H_0 satisfies

$$H_0{}^2 = [d(\ln a)/dt]\,|_{t\,=\,t_{now}} = 8\pi G\rho_{cr}/3 \equiv 1.17 \times 10^{-56} h^2 \text{ cm}^{-2} \qquad (10.2.1.2a)$$

or

$$H_0 = 1.08 \times 10^{-28} h \text{ cm}^{-1} \equiv 3.24 \times 10^{-18} h \text{ s}^{-1} \qquad (10.2.1.2b)$$

If we evaluate eq. 10.2.1.1 for the present time we find

$$k = [\Omega_\gamma + \Omega_m + \Omega_\Lambda - 1]H_0{}^2 \cong [\Omega_m + \Omega_\Lambda - 1]H_0{}^2 = (0.012^{+0.087}_{-0.082})H_0{}^2$$

$$= 6.10^{+44.2}_{-41.7} \times 10^{-59}\ \text{cm}^{-2} \qquad (10.2.1.2c)$$

Notice the error "bars" make the value of k uncertain. They result from the difficulties in experimentally measuring the densities Ω_m and Ω_Λ.

The range of values for k in eq. 10.2.1.2c is $[-47.8 \times 10^{-59}, 50.3 \times 10^{-59}]$. Therefore we feel that the radius value determined from WMAP data in eq. 10.1.3, even though it is a lower bound, may be a better indication of the value of k assuming a closed Robertson-Walker universe. In any case we will use this value for k in eq. 10.2.4. This value is only 25% different from the estimate in eq. 10.2.1.2c and thus well within the order of magnitude goal of our calculations. The value of k that we have selected:

$$k = 4.57 \times 10^{-59}\ \text{cm}^{-2} \qquad (10.2.4)$$

implies

$$\Omega_m + \Omega_\Lambda = 1.009$$

which is within the error bars of these quantities.

We now define

$$\xi = k/H_0^{\,2} = 3.92 \times 10^{-3} h^{-2} \qquad (10.2.1.3)$$

for later use.

The solution of eq. 10.2.1.1 can be put into the form of integrals representing combinations of elliptic integrals:

$$\int_{a(t')}^{a(t)} da\ a\ H_0^{-1}[\Omega_\Lambda a^4 - \xi a^2 + \Omega_m a + \Omega_\gamma]^{-½} = \int_{t'}^{t} dt \qquad (10.2.1.4)$$

The result of these integrations is an implicit equation for a(t) that cannot be expressed in a simple closed form in terms of known functions. This equation can be easily solved numerically. A graph of a(t) is displayed in Fig. 10.2.1.1. Although it looks linear there are significant non-linearities in various parts of the plot of a(t). The radiation-dominated phase is not visible. It is a small slice of the plot since it amounts to less than 10^{13} s.

Because we know physically that approximations are possible for each of the various epochs: the matter-dominated epoch, the radiation-dominated epoch and so on, we can find physically meaningful approximations for each epoch. We therefore provisionally divide the life of the universe into two epochs: an explosive growth epoch, and an expanding epoch subdivided into matter-dominated and radiation-dominated phases. We would have subdivided the expanding epoch into three phases if the constant ξ were not so small.

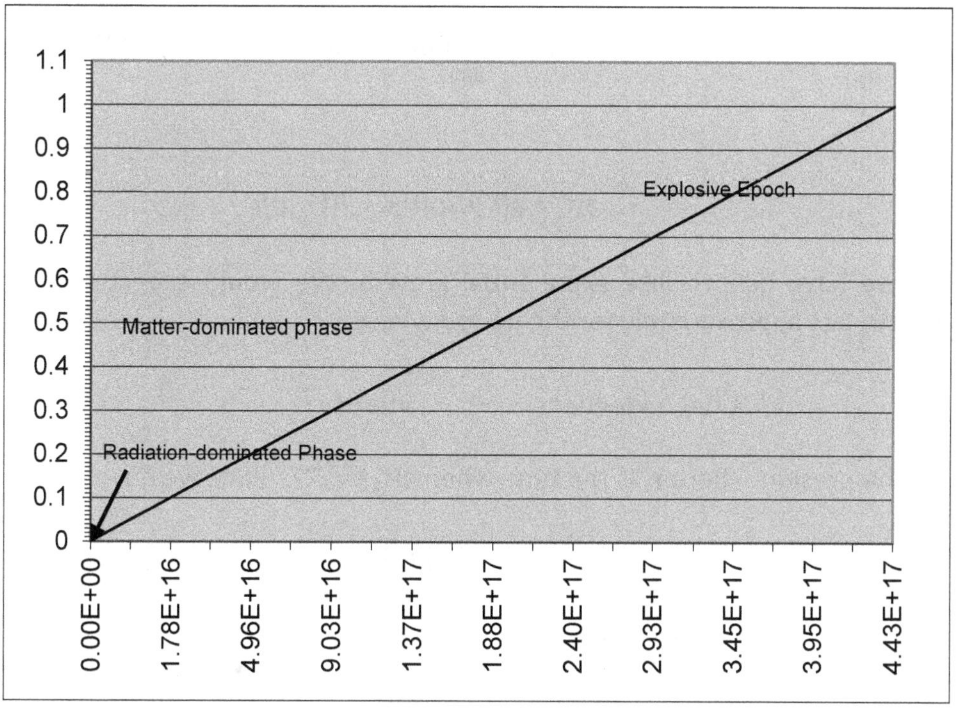

Figure 10.2.1.1. A plot of a(t) generated from eq. 10.2.1.4 through numerical integration.

Epoch	Type	Phases
I	**Explosive Growth**	
II	**Expanding**	**Matter-dominated** **Radiation-dominated**

Table 10.2.1.1. Epochs and phases of the universe after the Big Bang Epoch.

10.2.2 The Explosive Growth Epoch

The integral on the left in eq. 10.2.1.4 appears to support a simple approximation during the explosive growth phase. Notice that at $t = t_{now}$ we have $-\xi a^2 + \Omega_m a + \Omega_\gamma \cong .308$. The sum of these terms gets smaller as we proceed into the past. Therefore we approximate the left side with

$$\int_{a(t')}^{a(t)} da\, aH_0^{-1}[\Omega_\Lambda a^4]^{-1/2} = (H_0\Omega_\Lambda^{1/2})^{-1} \int_{a(t')}^{a(t)} da/a = (H_0\Omega_\Lambda^{1/2})^{-1}\ln[a(t)/a(t')] \cong t - t' \tag{10.2.2.1}$$

or

$$a(t) = a(t')\exp[H_0\Omega_\Lambda^{1/2}(t - t')] \tag{10.2.2.2}$$

Thus we have deSitter-like exponential growth. We would expect this growth phase to last approximately for the period where

$$\Omega_\Lambda a^4(t_E) \approx \Omega_m a(t_E) \quad \Rightarrow \quad a(t_E) = .77 \tag{10.2.2.3}$$

until the present where t_E is the time when $a(t_E) = .77$. Since we have normalized $a(t)$ by

$$a(t_{now}) = 1 \tag{10.2.2.4}$$

we see that eq. 10.2.2.2 requires

$$a(t) = \exp[H_0\Omega_\Lambda^{1/2}(t - t_{now})] \tag{10.2.2.5}$$

in the epoch that we have called the Explosive Growth Epoch. Note

$$H_0\Omega_\Lambda^{1/2} = 1.78 \times 10^{-18}\ s^{-1} \tag{10.2.2.6}$$

The beginning of this epoch is set by eq. 10.2.2.3:

$$t_E = t_{now} + (H_0\Omega_\Lambda^{1/2})^{-1}\ln(.77) = 2.99 \times 10^{17}\ s \tag{10.2.2.7}$$

giving the time interval of this epoch as 1.47×10^{17} s = 4.7 Gyr —much longer than the radiation-dominated era, and still expanding. The transition time t_E is close to the standard hypothesis for the appearance of Dark Energy of at a red-shift $z \sim .5$ or $a(t) = 2/3$. ($1 + z = a^{-1}$) The a(t) value of 2/3 corresponds to a time $t_E = 2.7 \times 10^{17}$ s by eq. 10.2.1.4 (numerical solution) which is to be compared to our estimate $t_E = 2.99 \times 10^{17}$ s – a roughly 10% difference.

10.2.3 The Expanding Epoch

The Expanding Epoch includes the matter-dominated and radiation-dominated phases. The solution for the scale factor in this epoch is obtained by neglecting the Ω_Λ term in eq. 10.2.1.4:

$$\int_{a(t')}^{a(t)} da\, aH_0^{-1}[-\xi a^2 + \Omega_m a + \Omega_\gamma]^{-1/2} \cong \int_{t'}^{t} dt \qquad (10.2.3.1)$$

This equation is easily integrated yielding:

$$[-\xi a^2(t) + \Omega_m a(t) + \Omega_\gamma]^{1/2} + \Omega_m(2\xi^{1/2})^{-1}\arcsin[(\Omega_m - 2\xi a(t))(\Omega_m^2 + 4\xi\Omega_\gamma)^{-1/2}] -$$

$$-[-\xi a^2(t') + \Omega_m a(t') + \Omega_\gamma]^{1/2} - \Omega_m(2\xi^{1/2})^{-1}\arcsin[(\Omega_m - 2\xi a(t'))(\Omega_m^2 + 4\xi\Omega_\gamma)^{-1/2}] =$$

$$= -\xi H_0(t - t') \qquad (10.2.3.2)$$

Assuming $a(0) = 0$ and letting $t_0 = 0$ we find

$$[-\xi a^2(t) + \Omega_m a(t) + \Omega_\gamma]^{1/2} + \Omega_m(2\xi^{1/2})^{-1}\arcsin[(\Omega_m - 2\xi a(t))(\Omega_m^2 + 4\xi\Omega_\gamma)^{-1/2}] -$$

$$- \Omega_\gamma^{1/2} - \Omega_m(2\xi^{1/2})^{-1}\arcsin[\Omega_m(\Omega_m^2 + 4\xi\Omega_\gamma)^{-1/2}] = -\xi H_0 t \qquad (10.2.3.3)$$

Eq. 10.2.3.3 can be substantially simplified. Note the arguments of the arcsines are both near one in value due to the smallness of Ω_γ and ξ. Both arcsines can be approximated using

$$\arcsin(1 - \epsilon) \cong \pi/2 - (2\epsilon)^{1/2} \qquad (10.2.3.4)$$

for small ϵ.

First we approximate the arguments of the arcsines with

$$\arcsin[(\Omega_m - 2\xi a(t))(\Omega_m^{\,2} + 4\xi\Omega_\gamma)^{-1/2}] \cong \arcsin(1 - 2\xi a(t)\Omega_m^{\,-1} - 2\xi\Omega_\gamma\Omega_m^{\,-2})$$

and

$$\arcsin[\Omega_m(\Omega_m^{\,2} + 4\xi\Omega_\gamma)^{-1/2}] \cong \arcsin(1 - 2\xi\Omega_\gamma\Omega_m^{\,-2})$$

Then using eq. 10.2.3.4 we obtain

$$[-\xi a^2(t) + \Omega_m a(t) + \Omega_\gamma]^{1/2} - [\Omega_m a(t) + \Omega_\gamma]^{1/2} \cong -\xi H_0 t \tag{10.2.3.5}$$

Noting that $\xi a^2(t)$ is much smaller than the other terms in the first square root in eq. 10.2.3.5 we can further approximate that equation by expanding the square root to obtain:

$$a^2(t)[\Omega_m a(t) + \Omega_\gamma]^{-1/2} \cong 2H_0 t \tag{10.2.3.6}$$

Eq. 10.2.3.6 embodies the standard matter-dominated and radiation-dominated expressions for the scale factor.

In the case of the matter-dominated phase we have

$$\Omega_m a(t) > \Omega_\gamma$$

and can approximate eq. 10.2.3.6 accordingly

$$a^2(t)[\Omega_m a(t)]^{-1/2} \cong 2H_0 t \tag{10.2.3.7}$$

or

<u>Matter-dominated Phase</u>

$$a(t) \cong [2\Omega_m^{\,1/2} H_0 t]^{2/3} \tag{10.2.3.8}$$

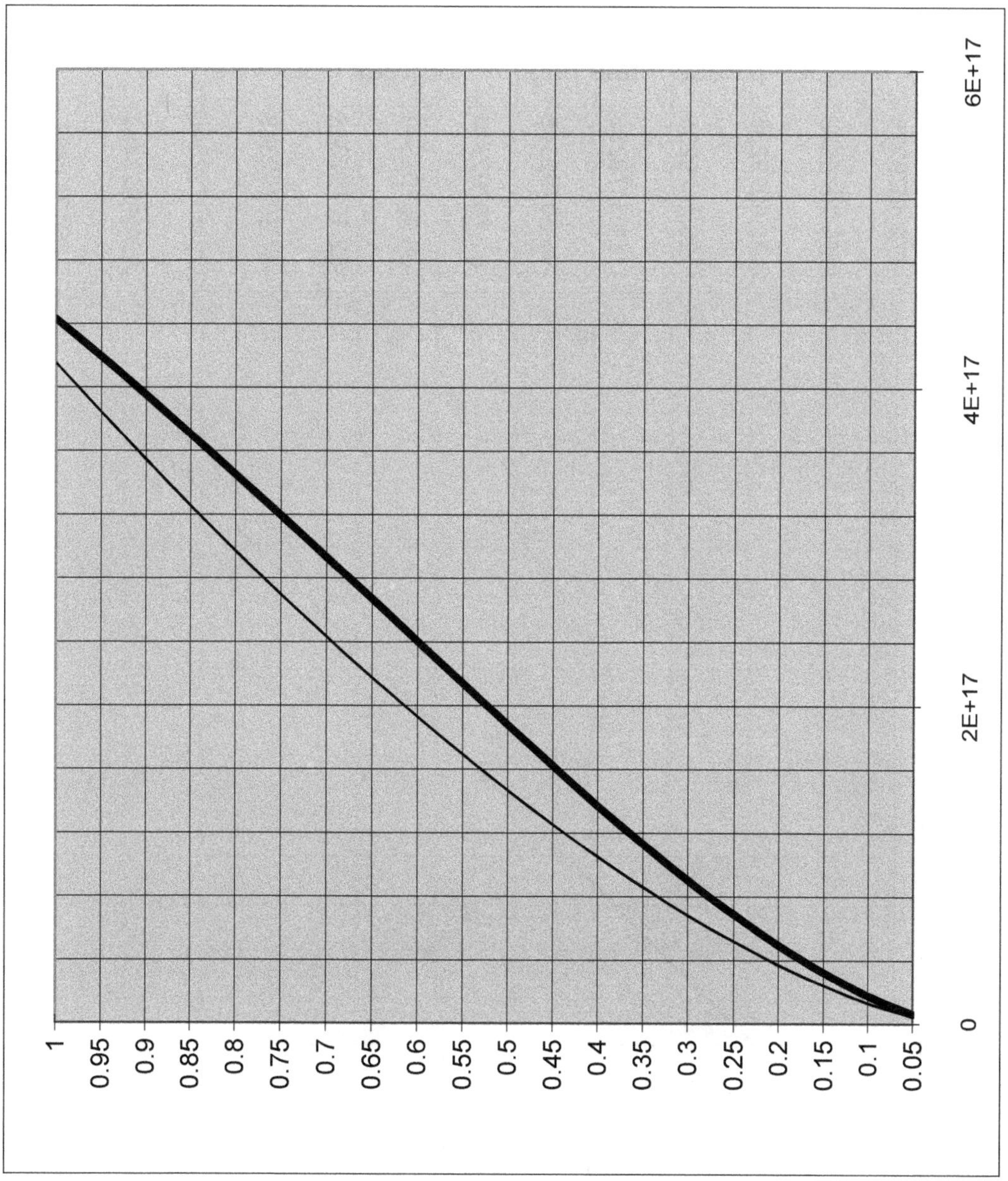

Figure 10.2.3.1. A plot of a(t) (horizontal axis) vs. time in seconds. The thick line is the plot of a(t) obtained by direct numerical integration of eq. 10.2.1.4 including the three density terms and the curvature constant term. The thin line is a plot of a(t) calculated directly from the approximation eq.

10.2.3.6. The approximation becomes increasingly better for small times as t → 0. (Reader: please rotate page 90 degrees clockwise.)

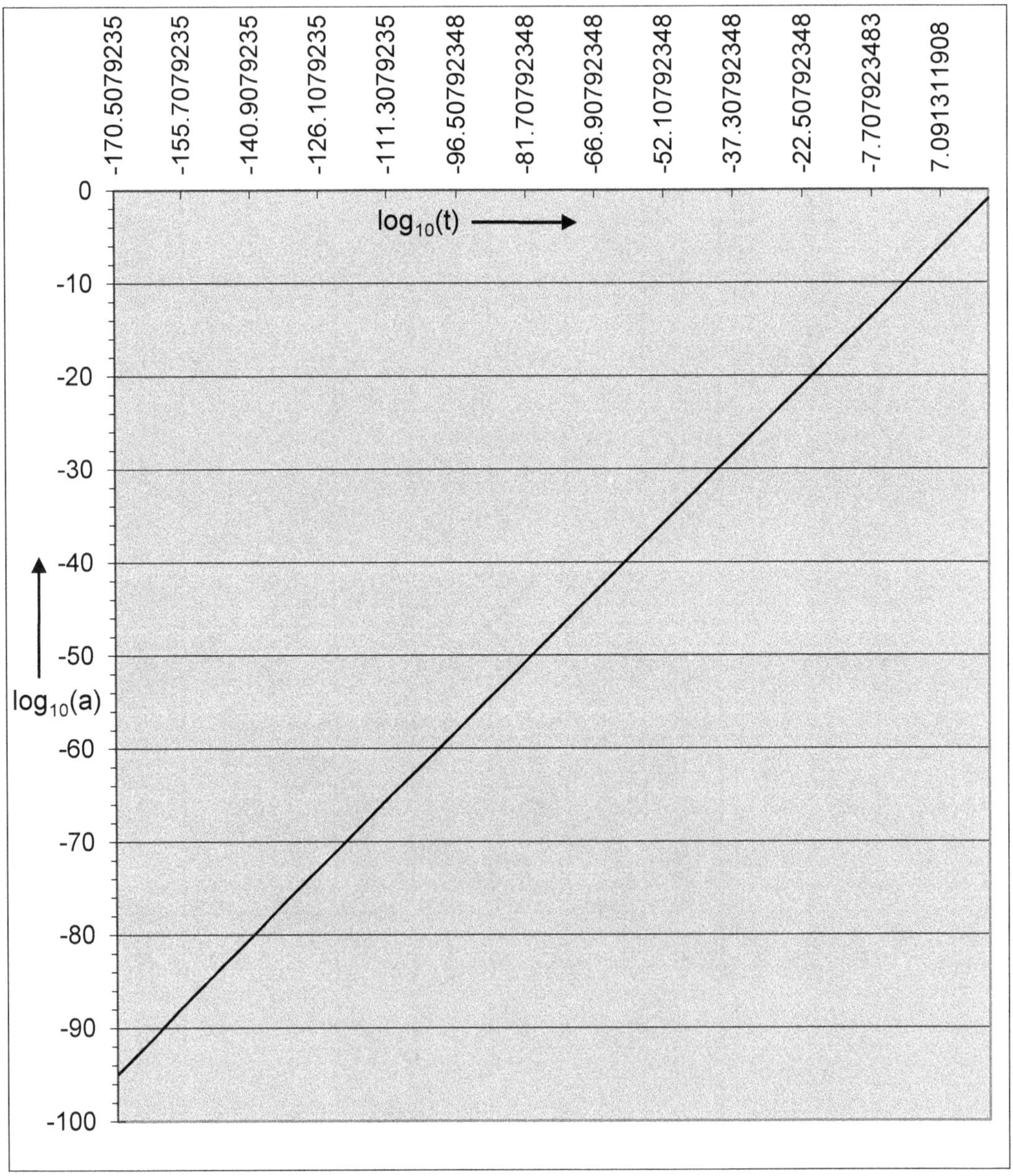

Figure 10.2.3.2. A log-log (base 10) plot of a(t) vs. t (in seconds) for small times calculated from eq. 10.2.3.6.

In the case of the radiation-dominated phase we have

$$\Omega_m a(t) < \Omega_\gamma$$

due to the smallness of the scale factor in that phase We approximate eq. 10.2.3.6 accordingly

$$a^2(t)[\Omega_\gamma]^{-½} \cong 2H_0 t \qquad (10.2.3.9)$$

Thus

Radiation-dominated Phase

$$a(t) \cong [2\Omega_\gamma^{½} H_0 t]^{½} \qquad (10.2.3.10)$$

The crossover point between the radiation-dominated and matter-dominated phase is at

$$\Omega_m a(t_{RM}) = \Omega_\gamma \qquad \text{or} \qquad a(t_{RM}) = 1.79 \times 10^{-4} \qquad (10.2.3.11)$$

where t_{RM} is the crossover time:

$$t_{RM} = 7 \times 10^{11}\ \text{s} \qquad (10.2.3.12)$$

Figs. 10.2.3.1 and 10.2.3.2 contain plots of a(t). Fig. 10.2.3.1 shows the approximate implicit equation for a(t) (eq. 10.2.3.6) is quite good over the entire range, and particularly good for small times in the radiation-dominated time frame. Since this region is the region of interest as it connects to the Big Bang Epoch we shall use this approximation, and eq. 10.2.3.10, for a(t) at very small times near t = 0.

10.3 Two-Tier Quantum Big Bang Model at the Beginning of Time

Section 10.2 presented the approximate expressions for the scale factor near t = 0. In this section we will consider numerical estimates for the scale factor, and other quantities of interest, such as the temperature and density near, and at, t = 0 in the Two-Tier model whose early time behavior is described in chapter 9.

In estimating quantities we are confronted with an imprecise determination of the needed input parameters because of experimental uncertainties and the impossibility of currently finding certain parameters experimentally in a model independent way. So we will use reasonable estimates for input parameters realizing that they may sometimes be off by up to a few orders of magnitude. Because of the vast differences in value between terms comprising the scale factors we will see a few orders of magnitude is often not a significant issue in determining the relative importance of terms.

The reader may notice slight differences in values due to rounding off numbers to three figures in the text while keeping values to 16 significant digits in the calculations. These differences can have a cumulative effect so we want to emphasize that our goal is order of magnitude accuracy.

The input data items are those of section 10.2 plus:

1. The current temperature of the Cosmic Microwave Background (CMB) radiation = 2.725 °K.

2. We assume $M_c = M_{Planck} = 1.22 \times 10^{28}$ ev since M_{Planck} is the only large mass intrinsic to the theory of gravitation and thus seemed to be a natural choice.

10.3.1 The CMB Temperature

From the current CMB temperature (T = 2.725 °K) we find

$$\kappa T = \kappa T_0/a(t_{now}) = \kappa T_0 = 2.34 \times 10^{-4} \text{ ev} \qquad (10.3.1.1)$$

For later use we define

$$\chi = \pi^{3/2}\kappa T_0/(2M_c) \cong 5.34 \times 10^{-32} \qquad (10.3.1.2)$$

10.3.2 The Generalized Robertson-Walker Scale Factor

A_{BB} is related to B_{BB} by eq. 9.86. Our approximation for the B_{BB} scale factor is:[102]

[102] Please note that the physical value of scale factors is the absolute value of the complex scale factors (obtained by use of Reality group transformations) appearing here and in the following discussions.

$$B_{BB}(t, y) \cong (1 + M_c^2y^2)^{-1}\{-i\varkappa M_c y + k^{-1/2}M_c a(t) + [\varkappa^2 - 2i\varkappa y k^{-1/2}M_c^2 a(t) + k^{-1}M_c^2 a^2(t)]^{1/2}\} \qquad (10.1.5)$$

If we let $y = M_c^{-1}$ then we can find B_{BB} "at the borders of the universe" where it simplifies to:

$$B_{BB}(t) = B_{BB}(t, M_c^{-1}) \cong \{-i\varkappa + \varpi a(t) + [\varkappa^2 - 2i\varkappa\varpi a(t) + \varpi^2 a^2(t)]^{1/2}\}/2 \qquad (10.3.2.1)$$

with

$$\varpi = k^{-1/2}M_c = 9.17 \times 10^{61} \qquad (10.3.2.2)$$

If t and a(t) are small,

$$A_{BB} = A_{BB}(t, \check{r}) = M_c^{-1}B_{BB}(t, \check{r}) \cong \beta_0(\check{r}) + \beta_1(\check{r})a(t) + \ldots \qquad (10.3.2.3)$$

$$\beta_0(\check{r}) = \varkappa(1 - i\check{r})/[M_c(1 + \check{r}^2)] \qquad (10.3.2.4)$$

$$\beta_1(\check{r}) = k^{-1/2}(1 - i\check{r})/(1 + \check{r}^2) \qquad (10.3.2.5)$$

We noted earlier that if we transformed our results back to Robertson-Walker coordinates we would have a modified scale factor $a = a_{BBRW}$ which we can continue to express in terms of $\check{r} = M_c y$.

$$a(t) \to (1 + M_c^2y^2)k^{1/2}A_{BB}(t, y)/2 \equiv a_{BBRW}(t, \check{r}) \qquad (9.5.2.4)$$

Thus

$$a_{BBRW}(t, \check{r}) = (1 + M_c^2y^2)(2k^{-1/2}M_c)^{-1}B_{BB}(t, y) \qquad (10.3.2.6)$$
$$= \tfrac{1}{2}\{a(t) - i\varkappa\varpi^{-1}\check{r} + [(\varkappa/\varpi)^2 + a^2(t) - 2i(\varkappa/\varpi)\check{r}a(t)]^{1/2}\} \qquad (10.3.2.7)$$

by eq. 9.92. If we evaluate a_{BBRW} at $\check{r} = M_c y = 1$ (the maximum value of y), since it determines the "size" of the universe, then

$$a_{BBRW}(t) \equiv a_{BBRW}(t, 1) = \{a(t) - i\gamma + [\gamma^2 + a^2(t) - 2i\gamma a(t)]^{1/2}\}/2 \qquad (10.3.2.8)$$

with the dimensionless constant

$$\gamma = \chi/\varpi = 5.82 \times 10^{-94} \quad (10.3.2.9)$$

This value reminds one of Eddington's famous remark that cosmological quantities often have orders of magnitude that are approximately multiples of 90.

The real and imaginary parts of $a_{BBRW}(t)$ are:

$$\text{Re } a_{BBRW}(t) = a(t)/2 + [R(t)(1 + \cos \psi(t))/2]^{½}/2 \quad (10.3.2.10)$$

and

$$\text{Im } a_{BBRW}(t) = -\gamma/2 - [R(t)(1 - \cos \psi(t))/2]^{½}/2 \quad (10.3.2.11)$$

where

$$R(t) = [(\gamma^2 + a^2(t))^2 + 4\gamma^2 a^2(t)]^{½} \quad (10.3.2.12)$$

and

$$\cos \psi(t) = (\gamma^2 + a^2(t))/R \quad (10.3.2.13)$$

There are 2 distinctly different periods specified by the time dependence of a_{BBRW}. The first period corresponds to a universe of "slowly" increasing size. The second period is the radiation and matter dominated phases with a fairly rapid increase in a(t). The boundary time t_c between these periods is specified by:

$$\gamma = a(t_c) \quad (10.3.2.14)$$

yielding

$$t_c = 1.05 \times 10^{-167} \text{ s} \cong 10^{-167} \text{ s} \quad (10.3.2.15)$$

In this period we find

$$\text{Re } a_{BBRW}(0) \cong \gamma/2 = 2.91 \times 10^{-94} \quad (10.3.2.16a)$$

$$\text{Re } a_{BBRW}(t_c) \cong 1.28\gamma = 7.45 \times 10^{-94} \quad (10.3.2.16b)$$

$$\text{Im } a_{BBRW}(0) = -\gamma/2 = -2.91 \times 10^{-94} \quad (10.3.2.16c)$$

with the radius, and volume, of the universe "slowly" increasing during this period. The nature of this period directly reflects the effects of the blackbody Y quanta. This can be seen from its inverse square dependence on M_C. As $M_C \to \infty$ the constant $\gamma \to 0$ and thus $a_{BBRW} \to 0$ with the universe scaling down to zero size with the attendant catastrophes of infinite density and temperature that appear in the standard models. The blackbody quanta (Dark Energy) give the universe a meta-stable initial size thus avoiding catastrophic divergences.

Epoch	Type	Phases	Time Period
I	**Explosive Growth**	**Dark Energy-dominated**	2.99×10^{17} s – 4.46×10^{17} s
II	**Expanding**	**Matter-dominated** **Radiation-dominated**	7×10^{11} s – 2.99×10^{17} s 1.05×10^{-167} s – 7×10^{11} s
III	**Metastable Big Bang**	**Blackbody Y quanta (Dark Energy) dominated**	0 s – 1.05×10^{-167} s

Table 10.3.2.1 Epochs and phases of the Universe since t = 0.

The period after t_c is dominated by the usual scale factor a(t). This scale factor shows up directly in the real part of a_{BBRW}. Its behavior is:

$t < t_c$ (or $a(t) < \gamma$)

$$\text{Re } a_{BBRW}(t) \cong \gamma/2 + a(t)/2 \qquad (10.3.2.17a)$$

$t_c < t < t_{now}$

$$\text{Re } a_{BBRW}(t) \cong a(t) \qquad (10.3.2.17b)$$

The behavior of the imaginary part of a_{BBRW} is also indirectly dominated by a(t) but in a much less dramatic way. The gradual growth of the imaginary part of a_{BBRW} is more or less indicated by the following three values:

$$\text{Im } a_{BBRW}(0) \cong -\gamma/2 = -2.91 \times 10^{-94} \qquad (10.3.2.16c)$$

$$\text{Im } a_{BBRW}(t_c) \cong -0.822\gamma = -4.78 \times 10^{-94} \qquad (10.3.2.18a)$$

$$\lim_{t \to \infty} \text{Im } a_{BBRW}(t) = -\gamma = -5.82 \times 10^{-94} \qquad (10.3.2.18b)$$

and also the behavior

$a(t) \ll \gamma$ (or $t < t_c$)

$$\text{Im } a_{BBRW}(t) \cong -\gamma/2 - a(t)/2 \qquad (10.3.2.18c)$$

$\gamma \ll a(t)$ (or $t_c < t$)

$$\text{Im } a_{BBRW}(t) \cong -\gamma + O([\gamma/a(t)]^2) \cong -\gamma \qquad (10.3.2.18d)$$

Both the real and imaginary parts of a_{BBRW} roughly double in the time period $[0, t_c]$ and thereafter we see a gradual increase of Im a_{BBRW} in absolute value from $-.5\gamma$ to $-\gamma$ over the lifetime of the universe.

After t_c the real part grows dramatically (eq. 10.3.2.17) while the imaginary part remains minute. The details of the interpretation of the behavior of the scale factor a_{BBRW} in the Big Bang Epoch $[0, t_c]$ will be explored in chapter 11. It suffices, for now, to say the universe in the period before t_c is in a meta-stable state of "slowly" growing size due to the dynamics of the blackbody Y quanta. At t_c the epoch of the expanding universe as we know it begins!

In differentiating between the real and imaginary parts of a_{BBRW} we must realize that the Reality group combines them into a single real quantity when scaling coordinates. Happily the small size of the imaginary part it can be neglected in most situations.

Before proceeding to describe physically interesting features of the early universe we note the radial dependence of the scale factor $a_{BBRW}(t, \check{r})$ in general:

$$a_{BBRW}(t, \check{r}) = \{-i\gamma\check{r} + a(t) + [\gamma^2 - 2i\gamma\check{r}a(t) + a^2(t)]^{1/2}\}/2 \qquad (10.3.2.19)$$

$$\text{Re } a_{BBRW}(t, \check{r}) = a(t)/2 + [R(t, \check{r})(1 + \cos\psi(t, \check{r})/2]^{1/2}/2 \qquad (10.3.2.20)$$

$$\text{Im } a_{BBRW}(t, \check{r}) = -\gamma\check{r}/2 - [R(t, \check{r})(1 - \cos\psi(t, \check{r}))/2]^{1/2}/2 \qquad (10.3.2.21)$$

where

$$R(t, \check{r}) = [(\gamma^2 + a^2(t))^2 + 4(\gamma\check{r}a(t))^2]^{1/2} \qquad (10.3.2.22)$$

and

$$\cos\psi(t, \check{r}) = (\gamma^2 + a^2(t))/R(t, \check{r}) \qquad (10.3.2.23)$$

We find that the scale factor $a_{BBRW}(t, \check{r})$ is approximated in various time periods to well within an order of magnitude by

$0 \le t < t_c$

$$\text{Re } a_{BBRW}(t, \check{r}) \cong \gamma/2 + a(t)/2 \qquad (10.3.2.24)$$

$$\text{Im } a_{BBRW}(t, \check{r}) \cong -\gamma\check{r}/2 - a(t)\check{r}/2 \qquad (10.3.2.25)$$

$t = t_c$

$$\text{Re } a_{BBRW}(t_c, \check{r}) = \{1 + [(1 + \check{r}^2)^{1/2} + 1]^{1/2}\}\gamma/2 \le 1.28\gamma = 7.45 \times 10^{-94}$$

$$0 \ge \text{Im } a_{BBRW}(t_c, \check{r}) = \{-\check{r} - [(1 + \check{r}^2)^{1/2} - 1]^{1/2}\}\gamma/2 \ge -0.822\gamma$$

$t_c < t$

$$\text{Re } a_{BBRW}(t, \check{r}) \cong a(t)\{1 + (1 + \check{r}^2)\gamma^2/4\} \cong a(t) \qquad (10.3.2.26)$$

$$\text{Im } a_{BBRW}(t, \check{r}) \cong -\gamma\check{r}/2 - [\check{r}^2 + \gamma^2/(4a^2(t))]^{1/2}\gamma/2 \qquad (10.3.2.27)$$

$t \to \infty$

$$\text{Re } a_{BBRW}(t, \check{r}) \cong a(t)\{1 + (1 + \check{r}^2)\gamma^2/4\} \cong a(t) \qquad (10.3.2.28)$$

$$\text{Im } a_{BBRW}(t, \check{r}) \cong -\gamma\check{r} = -5.82 \times 10^{-94}\check{r} \qquad (10.3.2.29)$$

10.3.3 The Temperature of the Early Universe in the Generalized Robertson-Walker Metric

We will now examine the temperature of the early universe based on the results of the preceding section. For t near t = 0, we note κT depends on the blackbody scale factor of the generalized Robertson-Walker metric:[103]

$$\kappa T = \kappa T_0/B_{BB}(t, y) \qquad (10.3.3.1)$$

and

$$B_{BB}(t, y) = 2k^{-1/2}M_c(1 + M_c^2y^2)^{-1}a_{BBRW}(t, \check{r}) \qquad (10.3.3.2)$$

by eq. 10.3.2.6 using the variable

$$\check{r} \equiv M_c y \qquad (10.3.3.3)$$

for convenience. Therefore from eqns. 10.3.2.24 – 10.3.2.27 we see

[103] Here again we remind the reader that complex values for radii, temperature and so on are transformed by the Reality group to real values – their absolute values.

$0 \le t < t_c$

$$\kappa T_< \equiv \kappa T(0 \le t < t_c) \cong 2\kappa T_0(1 + i\check{r})/\chi$$
$$= 8.76 \times 10^{27}(1 + i\check{r}) \text{ ev} \qquad (10.3.3.4)$$

$t_c < t$

$$\kappa T_> \equiv \kappa T(t_c < t)$$

$$\cong \kappa T_0(a(t) + i\gamma\check{r})/[2k^{-½}M_c(1 + \check{r}^2)^{-1}(a^2(t) + \gamma^2\check{r}^2)]$$

$$\cong \kappa T_0(1 + \check{r}^2)/[2k^{-½}M_c a(t)] = 1.28 \times 10^{-66}(1 + \check{r}^2)/a(t) \text{ ev} \qquad (10.3.3.5)$$

We note that the temperatures calculated in this section are in generalized Robertson-Walker coordinates (eq. 9.2.4.5) and not in Robertson-Walker coordinates.

Note also that the physical temperature is the absolute value of the complex temperature due to the use of the Reality group.

10.3.4 Consistency of Y_{BB} Approximation with the Resulting Temperature near $t = 0$

We now address the question of whether the values found for κT are consistent with the approximations made in chapter 9 in order to obtain an expression for Y_{BB}:

$$\cos(\omega\kappa Tt) \approx 1 \qquad (9.66)$$
$$J_1(\omega\kappa Ty) \approx \omega\kappa Ty/2 \qquad (9.67)$$

The first approximation is valid if $\kappa Tt \sim 0$ since the Planck distribution factor makes the largest contribution to the integral come from small ω. The values of $\kappa T_<$ and $\kappa T_>$ show that for larger times κTt is very small:

$$\kappa T_< t_c \le 1.98 \times 10^{-124}$$
$$\kappa T_> t \le (3.89 \times 10^{-51}\ s^{-1})t/a(t) \le (3.89 \times 10^{-51}\ s^{-1})t_{now}/a(t_{now}) = 1.74 \times 10^{-33}$$
$$(10.3.4.1)$$

The second approximation is valid if $\kappa Ty \ll 1$ since the Planck distribution factor again makes the largest contribution to the integral come from small ω.

We find the maximum values of κTy in the two time periods from eqns. 10.3.3.4 and 10.3.3.5 to be

$0 \leq t < t_c$

$$\text{MAX}(|\omega\kappa T_< y|) = |\omega\kappa T_< M_c^{-1}| = 1.02\omega \qquad (10.3.4.2)$$

which is small since the dominant part of the integration comes from small ω; and

$t_c < t$

$$\text{MAX}(|\omega\kappa T_> y|) = |\omega\kappa T_>(t_c) M_c^{-1}| = 2.10 \times 10^{-94}\omega/a(t)$$
$$\leq 2.10 \times 10^{-94}\omega/a(\mathbf{t_c}) = 0.361\omega \qquad (10.3.4.3)$$

which is also small. Thus the Bessel function power series expansion is well approximated by its first term:

$$J_1(\omega\kappa Ty) = (\omega\kappa Ty/2) \qquad (10.3.4.4)$$

We conclude our approximate calculation of A(t, y) is valid for all time and for the complete range of y values: $0 \leq y \leq M_c^{-1}$.

10.3.5 Plots of the Scale Factor from t = 0 to the Present

We will create several plots of the scale factor from t = 0 to the present time $t = 4.46 \times 10^{17}$ s for the maximum value of $y = M_c^{-1}$, which corresponds to the maximum Robertson-Walker radius coordinate value $r = k^{-½}$. The approximation that we have developed for B_{BB} has been justified for times between the Big Bang and the present.

In Fig. 10.3.5.1 we show a log – log plot of the real and imaginary parts of $a_{BBRW}(t)$ vs. t using base 10 logarithms for $t \in [10^{-200}, 10^{20}]$ seconds. In Fig. 10.3.5.2 we plot the real and imaginary parts of a_{BBRW} vs. time from t = 0 to $t = 1.2 \times 10^{-246}$ s. In Fig. 10.3.5.3 we plot the real part of a_{BBRW}, and the Robertson-Walker scale factor a(t), vs. time in seconds in the period around 10^{-167} s. It shows the rapidity of the transition of Re a_{BBRW} from slowly rising to rapidly rising with a(t).

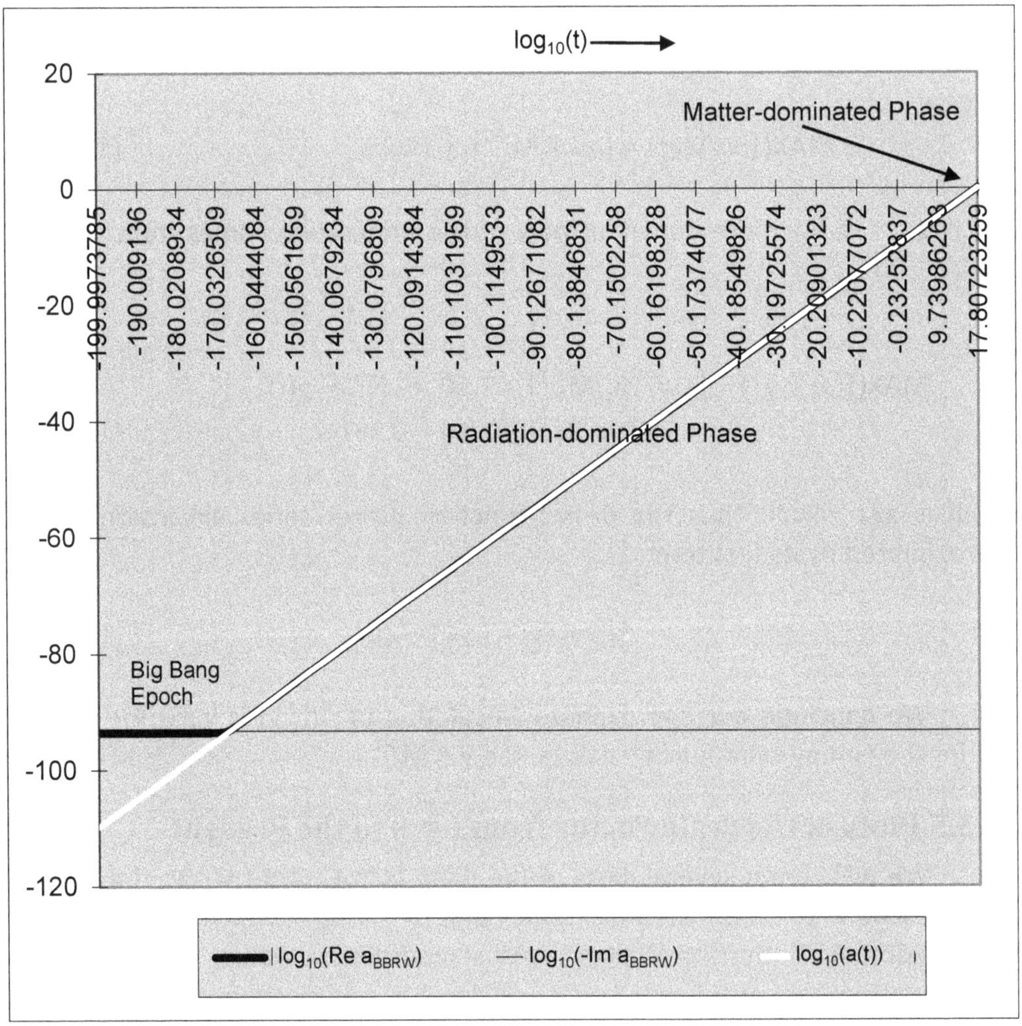

Figure 10.3.5.1. A log-log (base 10) plot of the real and imaginary parts of a_{BBRW} and the Robertson-Walker scale factor a(t) versus the log (base 10) of time in seconds. Note the imaginary part of a_{BBRW} is very slowly growing. The real part of a_{BBRW} is growing slowly until $t_c \approx 10–167$ s and thereafter equals a(t) to good approximation.

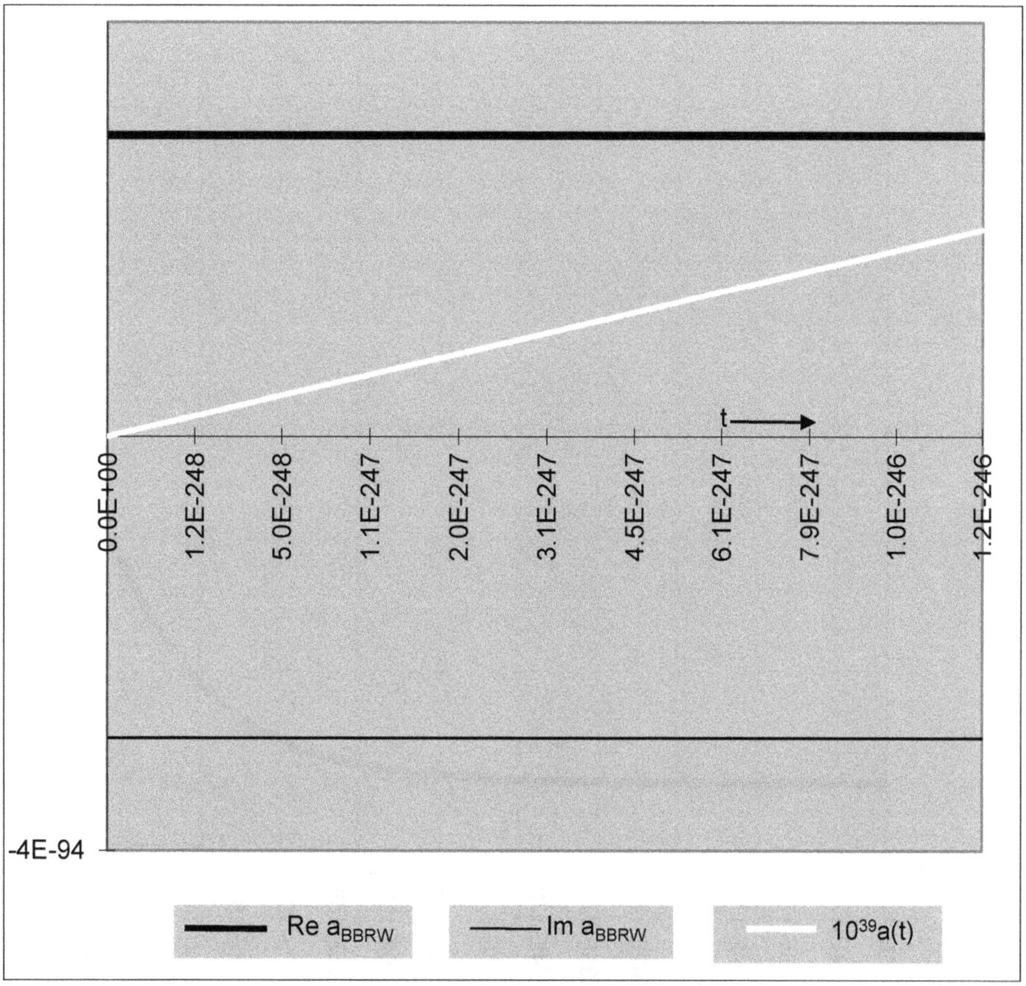

Figure 10.3.5.2. A plot of the real and imaginary parts of a_{BBRW}, and $10^{39} \times a(t)$, versus time from t = 0 to $t = 1.2 \times 10^{-246}$ s. Note they are slowly varying and well behaved in the neighborhood of t = 0 with only a(t) having the value of zero. "E" indicates a power of ten (for example: 2.0E-248 = 2.0×10^{-248} s).

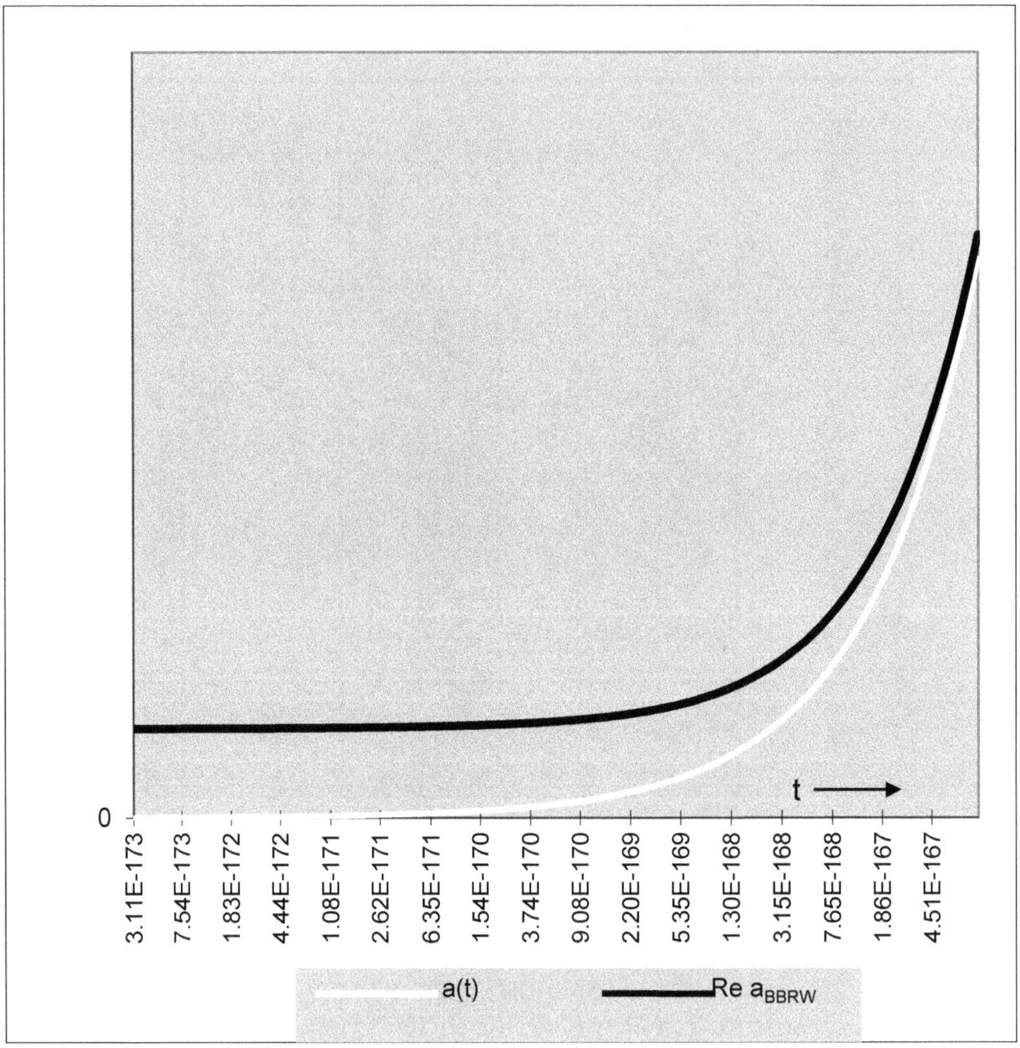

Figure 10.3.5.3. A plot of the real part of a_{BBRW} and a(t) vs. time in seconds around the time 10^{-167} s. Note Re a_{BBRW} quickly changes from slowly growing to growing like a(t).

10.4 The Interpretation of the Complex Scale Factor

The interpretation of the complex scale factor $a_{BBRW}(t, ř)$ hinges on its role in the expression for the proper interval. The expression for the proper interval in the Robertson-Walker metric is:

$$d\tau^2 = dt^2 - R^2(t)[dr^2/(1 - kr^2) + r^2(d\theta^2 + \sin^2\theta d\varphi^2)]$$

It was generalized to

$$d\tau^2 = dt^2 - A^2(t, ř)[dř^2 + ř^2(d\theta^2 + \sin^2\theta\, d\varphi^2)] \qquad (9\text{-}A.4.9)$$

and an identification was made between the Robertson-Walker scale factor R(t) ≡ a(t) and $a_{BBRW}(t, ř)$ through the following chain of equalities and correspondences

$$a(t) = a(t)b_O(r) = A(t, ř)/b(ř) \rightarrow A_{BB}(t, ř)(1 + ř^2)k^{1/2}/2 = a_{BBRW}(t, ř) \qquad (10.4.1)$$

where the → relates the classical expressions on the left with the expressions on the right that embody quantum corrections due to Y-quanta blackbody radiation. Eq. 10.4.1 is based on eqns. 9-A.4.6, 9-A.4.7, 9-A.4.9, 9-A.4.11, 9-A.4.12, and 9-A.6.4a. Combining eqns. 10.4.1 we find

$$d\tau^2 = dt^2 - a_{BBRW}(t, ř(r))^2[dr^2/(1 - kr^2) + r^2(d\theta^2 + \sin^2\theta d\varphi^2)] \qquad (10.4.2)$$

where the relation between ř and r, ř(r), is specified by eq. 9-A.4.10. We call the metric in eq. 10.4.2 a generalized Robertson-Walker metric. It is equivalent to eq. 9-A.4.9 – differing only in the definition of the radial coordinate.

Having determined the role of the complex scale factor in the metric tensor we can now physically interpret it by applying a Reality group transformation that effectively multiplies $a_{BBRW}(t, ř(r))$ by a phase making it real and equal to its absolute value $|a_{BBRW}(t, ř(r))|$.

We can simplify the physical interpretation without loss of generality by considering the radial coordinate of eq. 9-A.4.9 to be

$$r_{GRW} = A_{BB}ř \qquad (10.4.3)$$

$$= r_r + ir_i \tag{10.4.4}$$

where

$$A_{BB}(t, y) = 2a_{BBRW}(t, \check{r})/[(1 + M_c^2y^2)k^{½}] \tag{9.5.2.4}$$

Thus

$$r_{GRW} = 2a_{BBRW}(t, \check{r})\check{r}/[(1 + \check{r}^2)k^{½}] \tag{10.4.5}$$

Applying a Reality group transformation we obtain the physically measurable, real-valued radius

$$r_{GRWphysical} = 2|a_{BBRW}(t, \check{r})\check{r}/[(1 + \check{r}^2)k^{½}]| \tag{10.4.5a}$$

10.5 Time Evolution of the Hubble Rate

Hubble's rate is one of the linchpins of modern cosmology. It is determined by Einstein's equation eq. 10.2.1.1 when written in the form:

$$H(t) = \dot{a}/a = [H_0^2(\Omega_\gamma/a^4(t) + \Omega_m/a^3(t) + \Omega_\Lambda) - k/a^2(t)]^{½} \tag{10.5.1}$$

At small times, in the radiation-dominated phase, eq. 10.5.1 can be approximated by

$$H(t) \cong H_0\Omega_\gamma^{½}/a^2(t) \tag{10.5.2}$$

If we define a Hubble rate H(t) using $a_{BBRW}(t, \check{r})$ then

$$H_{BBRW}(t, \check{r}) = |\dot{a}_{BBRW}(t, \check{r})/a_{BBRW}(t, \check{r}) \equiv \dot{A}_{BBRW}(t, \check{r})/A_{BBRW}(t, \check{r})|$$

$$= |[H_0^2(\Omega_\gamma/a_{BBRW}^4(t, \check{r}) + \Omega_m/a_{BBRW}^3(t, \check{r}) + \Omega_\Lambda) - k/a_{BBRW}^2(t, \check{r})]^{½}| \tag{10.4.3}$$

$H_{BBRW}(t, \check{r})$ is the same as H(t) until we reach the first instants of the universe which we have called the Big Bag Epoch. Then we find

$0 \le t < t_c$

$$H_{BBRW}(t, \check{r}) \cong H_0\Omega_\gamma^{½}/|a_{BBRW}(t, \check{r})|^2 \tag{10.4.4}$$

where

$$\mathrm{Re}\ a_{BBRW}(t, \check{r}) \cong \gamma/2 + a(t)/2 \quad (10.3.2.24)$$

and

$$\mathrm{Im}\ a_{BBRW}(t, \check{r}) \cong -\gamma\check{r}/2 - a(t)\check{r}/2 \quad (10.3.2.25)$$

Substituting in eq. 10.4.4 we find

$$H_{BBRW}(t, \check{r}) \cong 4H_0\Omega_\gamma^{\frac{1}{2}}|[1 - \check{r}^2 + 2i\check{r}]/[(\gamma + a(t))^2(1 + \check{r}^2)^2]| \quad (10.4.5)$$

$|H_{BBRW}(t, \check{r})|$ is a real physical number in this range. Thus space has a Hubble rate that is both space and time dependent in the Big Bang Epoch.

At $t = 0$ we find $H_{BBRW}(0, \check{r})$ is finite unlike the radiation-dominated Hubble rate (eq. 10.5.2):

$$H_{BBRW}(0, \check{r}) \cong 4H_0\Omega_\gamma^{\frac{1}{2}}|[1 - \check{r}^2 + 2i\check{r}]/[\gamma^2(1 + \check{r}^2)^2]| \quad (10.4.6)$$

At the "edge" of the universe the Hubble rate is

$$H_{BBRW}(0, 1) \cong 2H_0\Omega_\gamma^{\frac{1}{2}}/\gamma^2 = 9.51 \times 10^{166}\ \mathrm{s}^{-1} \quad (10.4.7)$$

and is solely due to the imaginary part of $4H_0\Omega_\gamma^{\frac{1}{2}}[1 - \check{r}^2 + 2i\check{r}]/[(\gamma + a(t))^2(1 + \check{r}^2)^2]$ since the real part is zero.

Thus we consistently avoid the divergences that appear at $t = 0$ in the Standard Cosmological Model. Its radiation-dominated phase's Hubble rate is $(2t)^{-1}$ which diverges at $t = 0$.

11. The Big Bang Epoch

No great thing is created suddenly.
Discourses - *Epictetus*

11.1 The $t = 0$ Big Bang Scenario

The Two-Tier cosmological theory that we have developed in preceding chapters differs dramatically from the Standard Cosmological Model in the Big Bang Epoch and yet smoothly melds into the Standard Cosmological Model in the Expanding Universe and Exploding Universe epochs.

The universe has a finite size, temperature and density at the point of the Big Bang which we define to be at the time t = 0. The universe grows slowly for a period of time (until roughly 10^{-167} s) that we call the Big Bang Epoch. In this time period we see a very hot, very dense, macroscopic conglomeration of Y quanta, radiation and elementary particles that coexist with each other with non-singular interactions. These particles are not localized. Each particle can be said to occupy the entire universe since the size of the universe is infinitesimal compared to any particle's Compton radius (if it has one). The universe has an almost classical Robertson-Walker type of metric. In particular, it has a generalized Robertson-Walker metric with quantum corrections due to an effectively classical Y-quanta blackbody radiation field (the inflatons) that is both the source of the metastability of the universe at t = 0 and the source of infinitesimal imaginary spatial dimensions that are comparable in extent with the size of the real spatial dimensions of the universe during the Big Bang Epoch.

As the universe expands due to the Y quanta energy it appears that enormous amounts of gravitational energy are also released since the Two-Tier gravitational potential is zero at r = 0 and has a minimum around $r \approx 10^{-33}$ cm. Thus we view the universe in the Big Bang Epoch as in a "slowly" expanding metastable state. (This state is comparable to the metastable false vacuum state in inflation theories – but no scalar bosons are needed – the Y quanta play that role. The combination of gravitation and an effectively classical Y field serve to generate the metastable initial state of the Big Bang.

We can summarize our model's features (many of which are calculated later in this chapter) with:

1. The universe is a macroscopic object in terms of content and as such can be described by classical physics – modified by quantum effects.

2. Quantum fluctuations do not play any significant role in the Big Bang Epoch because of the nature of Two-Tier quantum field theory. For example, the quantum fluctuations of the quantized gravitational field were shown to be zero in chapter 7 of Blaha (2004):

$$<0|h_{\mu\nu}(X)h_{\alpha\beta}(X)|0> = \int d^3p\, b'_{\mu\nu\alpha\beta}(p)\, e^{-p^ip^j\Delta_{Tij}(0)}/[(2\pi)^3 2\omega_p] = 0 \qquad (7.3.8.3.3)$$

3. All Two-Tier quantum fields also have zero quantum fluctuations. This behavior holds whether they are quantized in flat space or in a curved space. Thus quantum fluctuations (or foam) are not an issue in Two-Tier theories.

4. The radiation and matter in the Big Bang Epoch produces, in effect, a classical Y-quanta blackbody spectrum that modifies the nature of space making it complex with the real and imaginary parts of space being comparable. This effect appears in the spatial scale factor of the generalized Robertson-Walker metric described in previous chapters. A Reality group transformation transforms spatial coordinates to real physical values.

5. The usual scale factor a(t) is determined by the conventional Standard Cosmological Model classical Einstein equation since its source is the macroscopic energy density.

6. The radius of the universe (with both real and imaginary parts) at the beginning of the Big Bang Epoch is

$$r_{universe}(0) = a_{BBRW}(0)/k^{1/2} = \gamma(1 - i)/(2k^{1/2}) = 4.30(1 - i) \times 10^{-65}\ \text{cm} \qquad (11.1.1)$$

The physical radius is the absolute value of $r_{universe}(0)$. The confinement of particles to a radius of this size means that they cannot be considered to be localized but, rather, they are spread over the entire volume of the universe.

7. The small size of the universe implies Two-Tier potentials between particles, and particle propagators, are effectively zero for all particles. Consequently the universe is in a metastable state. Some idea of the relative potential energy of Two-Tier QFT particles vs. ordinary QFT particles can be gleaned by comparing a standard Newton-Coulomb type of potential (knowing that it would be modified in strong gravitational fields if they were present)

$$V_{std} = 1/r \tag{11.1.2}$$

with the Two-Tier potential at short distances (below the Planck scale):

$$V_{tt} = 2\sqrt{\pi}\, M_c^2 r \tag{11.1.3}$$

obtained from eq. 7.3.9.3 of Blaha (2004). We have not displayed the coupling constant in eqns. 11.1.2 and 11.1.3. At r = $r_{universe}(0) = 4.30 \times 10^{-65}$ cm

$$V_{std} = 4.57 \times 10^{59} \text{ ev} \tag{11.1.4}$$

$$V_{tt} = 2\sqrt{\pi}\, M_c^2 r = 0.00115 \text{ ev} \tag{11.1.5}$$

Thus the Two-Tier potential between particles is negligible at distances up to the radius of the universe at t = 0 since the Heisenberg Uncertainty Principle, when applied using the "diameter" of the universe as the uncertainty in position, implies the uncertainty in a particle's energy, and thus the scale of particle energies, is (coincidentally) of the order of 10^{59} ev. (See the discussion in the following sections.)

8. In view of the above points it is reasonable, and self-consistent, to use the generalized Robertson-Walker metric that we have developed in the preceding chapters.

9. The t = 0 universe is very dense with an energy density of the order of 10^{339} g/cm^3 of particles with negligibly small, non-singular (as particle separation goes to zero) interactions. (See the discussion below.)

10. The Big Bang Epoch is a metastable state. Due to the form of the Two-Tier gravitational potential it has a much higher gravitational potential energy at t = 0 than when the radius of the universe is about 10^{-33} cm. **In fact, Gravity is a repulsive force (anti-gravity!) at distances less than 9.08 × 10^{-34} cm.** (Fig. 7.3.9.3 of Blaha (2004) is reproduced below for the reader's convenience.)

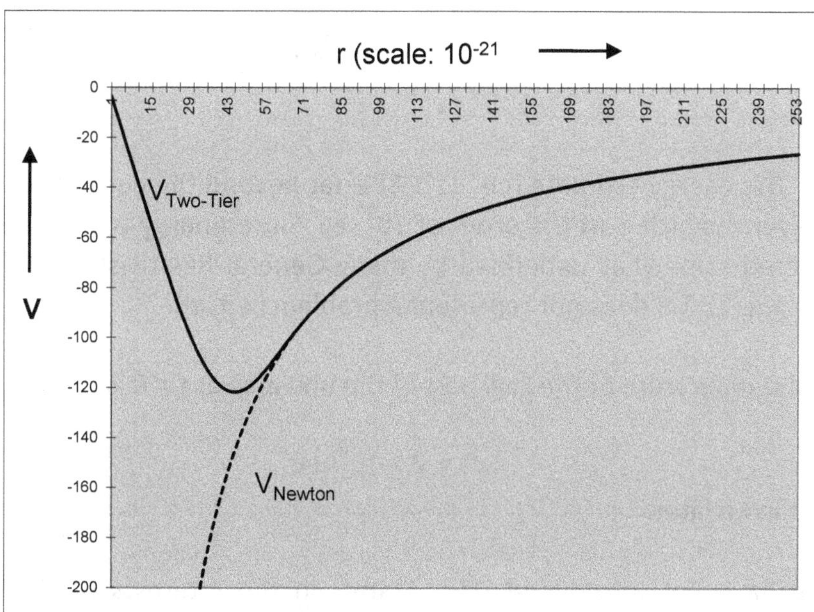

Figure 11.1.1. A plot of the Two-Tier gravitational potential (solid line) and the Newtonian gravitational potential (dashed line). Note anti-gravity at short distances. See Fig. 7.3.9.3 of Blaha (2004) for more details.

The Two-Tier gravitational potential between two particles has a minimum at (eq. 7.3.9.5 of Blaha (2004))

$$r_{MIN} = \pi^{-1/2} M_c^{-1} = 9.08 \times 10^{-34} \text{ cm} \tag{11.1.6}$$

At the minimum the gravitational $V_{Two\text{-}Tier}$ has the value:

$$V_{Two\text{-}TierMIN} = -.8427\sqrt{\pi}\, GM_c = -1.22 \times 10^{-28}\ ev^{-1} \qquad (11.1.7)$$

Since the energy of the universe is confined within a radius of 10^{-65} cm at t = 0 there is a tremendous release of gravitational potential energy as the universe expands to 10^{-33} cm and beyond. This energy is converted initially into kinetic energy and can be viewed as helping fuel the expansion. A crude approximation to this released energy is

$$\begin{aligned}\Delta E &\sim (\rho V_{universe})^2\, V_{Two\text{-}TierMIN} \\ &\cong (1.6 \times 10^{212}\ g/cm^3 \times 1.8 \times 10^{-100}\ cm^3)^2 \times 1.22 \times 10^{-28}\ ev^{-1} \\ &\cong 3 \times 10^{262}\ ev \qquad (11.1.8)\end{aligned}$$

The energy estimate (eq. 11.1.8) is far beyond the current total energy of the universe which is of the order of 10^{57} ev. Since energy is not conserved (It is considered somewhat undefined by many General Relativists.) as the universe evolves eq. 11.1.8 does not represent a problem in itself.

11. The temperature of the real part of the universe at t = 0 is

$$\kappa T \approx 2 \times 10^{89}\ ev \qquad (11.1.9)$$

as shown later.

We have not introduced Dark Matter in the interests of simplicity. Dark Matter would change the values of parameters but would not change the mechanism and the evolution of the universe qualitatively.

11.2 The Radius of the Universe in the Big Bang Epoch

In this section we will consider the radius of the universe at the point of the Big Bang (t = 0) and its relatively slow growth during the Big Bang Epoch ($t \leq t_c \cong 10^{-167}$ s).

The current radius of the universe (according to WMAP[104] data) is > 7.4 × 10^{28} cm. Motivated by that finding we have assumed the current radius of the universe in the Robertson-Walker model is twice that value:

Assumption: $$r_{universe}(t_{now}) = a(t_{now})/k^{1/2} \approx 2 \times 7.4 \times 10^{28}\ \text{cm} \quad (11.2.1)$$

and used it to determine k (eq. 10.2.4).

It is reasonable to define the radius of the universe at earlier times correspondingly. In our Two-Tier blackbody model for the universe it is

$$r_{universe}(t) = a_{BBRW}(t, \check{r} = 1)/k^{1/2} \quad (11.2.2)$$

with

$$a_{BBRW}(t, \check{r}) = \{-i\gamma\check{r} + a(t) + [\gamma^2 - 2i\gamma\check{r}a(t) + a^2(t)]^{1/2}\}/2 \quad (10.3.2.19)$$

Thus the radius of the universe is complex and given by

$$r_{universeBBRW}(t) = \{-i\gamma + a(t) + [\gamma^2 - 2i\gamma a(t) + a^2(t)]^{1/2}\}/(2k^{1/2}) \quad (11.2.3)$$

The physical radius of the universe is the absolute value of eq. 11.2.3. It is calculated by applying a Reality group transformation.

The real and imaginary parts of $r_{universeBBRW}(t)$ are:

$$\text{Re}\ r_{universeBBRW}(t) = \{a(t) + [R(1 + \cos\psi)/2]^{1/2}\}/(2k^{1/2}) \quad (11.2.4a)$$

and

$$\text{Im}\ r_{universeBBRW}(t) = \{-\gamma - [R(1 - \cos\psi)/2]^{1/2}\}/(2k^{1/2}) \quad (11.2.4b)$$

where

$$R = [(\gamma^2 + a^2(t))^2 + 4\gamma^2 a^2(t)]^{1/2} \quad (10.3.2.12)$$

and

$$\cos\psi = (\gamma^2 + a^2(t))/R \quad (10.3.2.13)$$

From the behavior of $a_{BBRW}(t)$ displayed in Figs. 10.3.5.1 – 10.3.5.3 we see that it is well approximated by eqns. 10.3.2.24 – 10.3.2.29 with $\check{r} = 1$. Therefore for $t < t_c$

[104] N. J. Cornish, D. N. Spergel, G. D. Starkman, and E. Komatsu, Phys. Rev. Lett. **92**, 201302-1 (2004).

$t < t_c$

$$\text{Re } r_{\text{universeBBRW}}(0 \leq t < t_c) \cong (\gamma + a(t))/(2k^{½}) \quad (11.2.5a)$$

$$\text{Re } r_{\text{universeBBRW}}(0) \cong \gamma/(2k^{½}) = 4.30 \times 10^{-65} \text{ cm} \quad (11.2.5b)$$

$$\text{Im } r_{\text{universeBBRW}}(0 \leq t < t_c) \cong -(\gamma + a(t))/(2k^{½}) \quad (11.2.6a)$$

$$\text{Im } r_{\text{universeBBRW}}(0) \cong -\gamma/(2k^{½}) = -4.30 \times 10^{-65} \text{ cm} \quad (11.2.6b)$$

from eqns. 10.3.2.24 and 10.3.2.25. Since the Planck length is 1.61×10^{-33} cm we see the real part of the radius of the universe at $t = 0$ (until $t \approx t_c$) is over thirty orders of magnitude smaller than the Planck length.

If we use eqns. 10.3.2.28 and 10.3.2.29 then the radius at present is

$t = t_{now}$

$$r_{\text{universeBBRW}}(t_{now}) = a(t_{now})/k^{½} - i\gamma/k^{½} \quad (11.2.7)$$

with

$$\text{Re } r_{\text{universeBBRW}}(t_{now}) = a(t_{now})/k^{½} = 1.48 \times 10^{29} \text{ cm} \quad (11.2.8)$$

as above (eq. 11.2.1), and

$$\text{Im } r_{\text{universeBBRW}}(t_{now}) = -\gamma/k^{½} = -8.60 \times 10^{-65} \text{ cm} \quad (11.2.9)$$

The ratio of the current, and the $t = 0$ real parts of the, radius of the universe is huge:

$$r_{\text{universe}}(t_{now})/(\text{Re } r_{\text{universeBBRW}}(0)) = 3.44 \times 10^{93} \quad (11.2.10)$$

The real part of the universe has expanded dramatically while the imaginary part has remained almost constant in size. The physical radius of the universe for $t < t_c$ and $t = t_{now}$ are the absolute values of the complex radius values above.

The value of the Robertson-Walker radius coordinate at points *within* the universe is specified by

$$r_{RW}(t, \check{r}) = a_{BBRW}(t, \check{r})r \tag{11.2.11}$$

The coordinate r is related to the ř coordinate by

$$r = 2k^{-½}\check{r}(1 + \check{r}^2)^{-1} \tag{9-A.4.11}$$

Thus

$$r_{RW}(t, \check{r}) = 2k^{-½}\check{r}a_{BBRW}(t, \check{r})(1 + \check{r}^2)^{-1} \tag{11.2.12}$$

The real and imaginary parts of $a_{BBRW}(t, \check{r})$ are specified in eqns. 10.3.2.24 – 10.3.2.29. Again we note the absolute values of eqns. 11.2.11-11.2.12 are the physical values of the Robertson-Walker radius coordinate.

11.2.1 Localization of Particles at t = 0

The radius of the universe in the neighborhood of t = 0 is approximately 6.08×10^{-65} cm. The particles within that incredibly small universe are still the "wave- particles" that we are familiar with within the framework of quantum mechanics. As such, the position and momentum of the particles must satisfy the Heisenberg Uncertainty Condition:

$$\Delta p \Delta x \geq \hbar \tag{11.2.1.1}$$

where $\hbar$ is Planck's constant divided by 2π. In view of the extraordinary small size of the universe – much smaller than the Compton wavelength of any known massive particle – the "spread" (uncertainty) in a particle's position is set by the radius of the universe:

$$\Delta x \approx 2\ |r_{universeBBRW}| \tag{11.2.1.2}$$

Thus the "spread" in the momentum of a particle (the "size" of the region in momentum space where the Fourier transform of the particle's wave function is large) is

$$\Delta p \approx \hbar/\Delta x \approx \hbar/[2|\mathrm{Re}\ r_{universeBBRW}|] = 1.62 \times 10^{59}\ \mathrm{ev} = 1.32 \times 10^{31} M_{Planck} \tag{11.2.1.3}$$

One can only view the particles in the universe in the neighborhood of t = 0 as spread across the entire universe. They are entirely unlocalized within the universe from a quantum viewpoint. Since all forces are non-singular in the very small universe at the beginning of time we can view the particles as co-resident in the same spatial region that constitutes the universe. (Simply put, they interpenetrate each other.) Thus the question of particle horizons and the homogeneity of the universe in the Beginning are irrelevant. *The universe today does not scale down to a micro-universe of the same sort as our universe in the neighborhood of t = 0.*

11.3 A Quantum Big Bang and Evolutionary Theory

At this point we have shown that our Quantum Big Bang theory does not have a singularity at the beginning of the universe and exhibits the known behavior of the universe since 350,000 years after the Big Bang epoch. Thus the long sought inflaton field turns out to be the quantum field, $Y^{\mu}(y)$, appearing in our definition of quantum coordinates. Remarkably our quantum coordinates also eliminate the divergences that have plagued quantum field theory for almost eighty years and enable calculations in The Standard Model and Quantum Gravity to be divergence free without the cumbersome renormalization techniques that were needed for ElectroWeak theory and would not work for Quantum Gravity.

So our quantum coordinates free The Standard Model (with or without our extension) and Quantum Gravity of infinities both now and at the beginning of the universe. Naturally this happy removal of infinities through one simple mechanism is a remarkable result that, we believe, reflects the simplicity of Nature when properly understood.

Appendix A. Asynchronous Logic as the Source of Two 4-dimensional Space-times

This appendix contains a shortened version of chapter 14 of Blaha (2011c). It provides a deep foundation for the existence a Flatverse containing two 4-dimensional surfaces (subspaces) – one being our universe; the other being a sister universe.

A.1 Asynchronous Logic Applied to Particles - From 4-Valued Logic to Dirac-like Equations and Four Generations

It is difficult to discern the truly fundamental general principles that give rise to The Standard Model. One new principle that this author feels is implicit in The Standard Model is a Principle of Asynchronicity. When processes take place in parallel whether it is Quantum Mechanical entanglement of processes at large distances from each other or in high order Feynman diagrams (or their old fashioned time ordered perturbation theory predecessor) the synchronicity of the processes is an issue. It is resolved by physical law that excludes incompatible asynchronous series of events. Situations do not arise where parallel processes get "out of sync" resulting in the failure of an entire physical process to execute properly.

In computation, asynchronicity issues may arise if the computation design is not properly done. For example parallel computations or computer processes on a chip or set of chips have to be carefully managed for a parallel computer process to complete properly. In the case of computer chip design (VLSI[105] chips and so on) techniques have been developed for the design of chips based on multi-valued logic. One conceptual approach uses a 4-valued logic to define clock-less computer logic circuits. The 4-valued logic developed by Fant (2005) has the four logic values TRUE, FALSE, NULL, and INTERMEDIATE. It is an extension of Boolean Logic that can accommodate asynchronicities in asynchronous computer circuits. It enables circuits to avoid the use of system

[105] An acronym for Very Large Scale Integration.

clocks to implement synchronization.[106] Thus the coordination is explicitly handled by a 4-valued logic and non-logical constructs are not needed. Concurrent transitions are coordinated solely by logical relationships with no need for any time constraint equations or relationships.[107]

The Standard Model, and all quantum field theories as well as Quantum Mechanics, contain asynchronicities. The structure of the particles and their dynamic equations prevent inconsistent physical behavior.

We suggest that a Principle of Asynchronicity is implicitly embodied in The Standard Model, and other quantum physics theories. This principle has major consequences in two areas: 1) it leads to Dirac-like equations for the fundamental fermions – leptons and quarks; and 2) it implies four generations of fermions due to a second 4-dimensional universe.

A.2 Features of 4-Valued Asynchronous Logic

In this section we will briefly consider its major features. For a detailed discussion see Fant (2005). The definitions of asynchronous circuits and asynchronous Logic are:

1. An *asynchronous circuit* is a circuit in which the component parts are autonomous and can act in parallel at various separations and various rates of time evolution. These circuits are not controlled by a clock mechanism but they do proceed or wait for signals indicating that they can proceed based on 4-valued logic.

2. *Asynchronous Logic* is the logic used in the design of asynchronous circuits. This logic embodies the asynchronicity and so circuits built using it do not use a clock to control the execution speed of the various parts of an asynchronous circuit. Consequently logic elements do not necessarily have a distinct true or false state at any given point in time. The logic supports "stop and go" states within an executing asynchronous circuit.

---

[106] Remarkably Bjorken (1965) pp 220-226 presents an analogy to VLSI designs, namely, Feynman diagrams treated as electrical circuits where momenta are mapped to currents, coordinates to voltages, Feynman parameters to resistance, and free particle equations of motion to Ohm's Law plus the equivalent of Kirchhoff's Laws. Thus Feynman diagrams and VLSI computer circuits are conceptually analogous. Particles interactions are asynchronous in space and time occurring at differing locations and times.

[107] If one wishes to use 2-valued logic in VLSI design then an additional formula is needed to maintain synchronization of the separate parts of a VLSI circuit.

In Fant's asynchronous 4-valued logic the four possible truth values are:

True – status is true and all data is current
False – status is false and all data is current
Intermediate – status is indefinite with some data current
NULL – status is indefinite with no data present – results in a suspension of processing of the circuit part in a NULL state until current data becomes present

"Data" is the information flowing through all or part of a circuit. Using these truth values the evolution in time of the parts of an asynchronous circuit are effectively synchronized by the logic without the use of a clock mechanism. A clock mechanism effectively is a subsidiary time constraint or set of time constraints that would be required in a 2-valued logic design. See Fant (2005) for further details.

An implicit aspect of asynchronous logic is the coordination of spatially separated parts of a circuit. Since spatial separations in a circuit can be mapped to time delays using the speed of data propagation between parts, spatial asynchronicites are normally subsumed under time asynchronicities. This is particularly true for computer chips which are kept small to minimize delays and maximize speed.

In the case of physical phenomena spatial separations cannot be transformed unambiguously into time delays since propagation speeds range from infinite (for quantum entanglements) to varying finite values. Thus in particle physics processes spatial asynchronicity is different in principle, and practice, from time asynchronicity.

A.3 A Physical Principle of Spatial and Time Asynchronicity

An obvious, and thus little thought of, feature of elementary particle phenomena is the coordination of the parts of a physical process in time and space. Complex Feynman diagrams embody the coordination of the spatially separated parts of interacting particles over a period of time. Quantum entanglement embodies the coordination of the parts of a physical phenomenon separated by small and large distances. These examples, which could be multiplied indefinitely, lead to a Physical Principle of Spatial and Time Asynchronicity.

Principle: Nature requires time and space asynchronicity. This asynchronicity is coordinated by 4-valued physico-logical structures for matter (fermions).

Elaboration: Elementary particle physical phenomena must support extended coordinated physical phenomena in space and time. The fundamental laws of particle physics must be such as to permit coordinated physical phenomena with coordination between their parts at large distances and for long time intervals. The coordination must be embodied within physical laws.

This principle will be shown to justify Dirac-like equations for particle dynamics, and to justify a four generation fermion spectrum based on a sister 4-dimensional universe. The origin of the fermion generations has been obscure since its discovery in the 1970's. This new principle furnishes a general basis for their origin.

Coordination is an obvious feature of physical phenomena. This principle embodies the obvious and shows it has significant physical implications. For example, if particles exist, then their antiparticles must also exist to support asynchronous behavior in interaction regions.[108] If only particles existed, then all interactions would proceed forward in time and the state of the interaction at any point in time would be determined. With the addition of antiparticles, asynchronicity is introduced and at various time slices of an interaction, the state may be ambiguous since antiparticles are negative energy particles moving backward in time.

A.4 Time Asynchronicity – Dirac-like Equations with Particles and Antiparticles

Time asynchronicities are common in the many subcircuits of a computer chip. Time asynchronicities are also common in the many interaction subregions of a set of particles in interaction. Fant (2005) has a diagram on p. 7 of a circuit with a set of subcircuits with five time slices of the interacting subcircuits showing five states of the "'data' wavefront" at five points in time. This diagram is similar to the time sliced diagram of an interacting system of particles in "old fashioned" time-ordered perturbation theory. Blaha (2005b) p. 29 displays such

---

[108] This comment applies to Majorana particles also which are their own anti-particles and thus capable of forward and backward motion in time.

a diagram (Fig. 5.1.4) in a description of a Standard Model Quantum Langauge Grammar – a language representation of particle physics. The diagram also appears in appendix B. Our diagram is similar to Fant's diagram[109] in its overall features as one might expect since both address time asynchronicity.

The asynchronicity that appears in perturbation theory diagrams is intimately related to the appearance of antiparticles in diagrams. As noted earlier antiparticles are interpretable as negative energy particles traveling backwards in time. The time orderings which are implicit in the Feynman diagram approach, and explicit in old fashioned perturbation theory, show the time asynchronicity, and the effect of the dynamics in coordinating the asynchronicities so that meaningful results follow from perturbative calculations.

Thus we are led to:

1. Fermion (matter) particles have four fundamental states corresponding to 4-valued Logic states
2. They are spin ½ particles in 4 dimensions.[110] A four state vector is a spinor in 4 dimensions.
3. They have 4×4 dynamical equations.

The form of the invariant distance expression implies that transformations between inertial reference frames must be elements in the complex 4-dimensional Lorentz group.

Interestingly, our association of particle states with truth values complements recent efforts to use particle spins, up and down, as storage for true and false in advanced computer devices. Complex particle diagrams thus are also interpretable as symbolic Logic computations.[111]

We conclude Time Asynchronicity leads to a complex 4-dimensional space-time. This space-time can be embedded in an 8-dimensional complex surface within the Flatverse.

---

[109] Due to copyright law we cannot show Fant's diagram.

[110] Weinberg (1995) p. 216 exhibits an equation that relates the number of components of a spinor to the dimension of its space-time. This equation shows that our 4 component vector when viewed as a spinor (under a mapping of logic values to spinor components) implies a 4-dimensional real or complex space-time. We choose a complex space-time in order to support superluminal transformations and tachyon particles.

[111] Blaha (2005).

A.5 Spatial Asynchronicity – Four Generations with Particle Oscillations

The Principle of Asynchronicity defined above also states the existence and synchronization of spatial asynchronicities. An excellent example of spatial asynchronicities is the flow of solar neutrinos to the earth. After much effort experimenters have found that solar neutrinos oscillate between electron type neutrinos and muon type neutrinos during their travel from the sun. The spatial oscillations of the stream of solar neutrinos between generations is a form of asynchronous behavior. On the basis of numerous such examples we find:

1. There are four levels – generations – of each fermion particle type corresponding to the four values in an asynchronous 4-valued logic for spatial asynchronicity. Just as in the case of time asynchronicity they lead to a map from 4 component logic vectors to spinors in a 4-dimensional universe. We have called this universe the sister universe earlier in this book and used it to understand the Higgs particle masses and the fundamental origin of mass and inertia.
2. Transitions between generations can occur asynchronously in streams of particles. These transitions and the corresponding states at various points along the path of the stream are coordinated by the fundamental laws of physics embodied in The Standard Model.

Thus a major unexplained feature of the fermion species – generations – is explainable by the spatial Principle of Asynchronicity upon the introduction of another 4-dimensional universe. We take this universe to be complex and embedded within 8 complex dimensions of the Flatverse.

A.6 A Deeper Basis for the Flatverse Containing Our Universe and a Sister Universe

Time and spatial asynchronicity are different in the case of particles because massive particle propagators do not have a fixed relation between time and distance in their propagation. In VLSI circuits the time interval between nodes is fixed by the distance between the nodes since the "wire" connecting them determines the speed of transmission. Thus we need two universes for particle asynchronicity but only one "universe" of discourse for VLSI circuits.

We need two 8-dimensional complex universes (surfaces) in the Flatverse. Their origin in Asynchronous Logic provides a much deeper understanding of the origin of the structure of the Flatverse.

Blaha (2005) provides more detail including a derivation of Dirac-like equations from matrix representations of 4-valued logic.

Appendix B. A Map between Interacting Particles and Computer Languages

In appendix A we showed that the origin of the Flatverse, containing two complex 8-dimensional subspaces (surfaces), could be derived from a Principle of Asynchronous Logic for elementary particles and physical processes. That discussion presumed that one could map the four components of a fermion spinor to logic values of a 4-valued Logic matrix representation.

In this appendix we will provide detailed support for the view that The Standard Model is a particular implementation of a general specification for a computation. In this view particles are skeletonized to data and interactions to the execution of a computation. Thus we provide a complementary view to the current effort to build quantum computers from atoms and molecules using Quantum Theory. The discussion in this chapter is largely condensed extracts of parts of Blaha (2005b). Our earlier works, beginning with Blaha (1998), contain the basic ideas and their elaboration.

B.1 The Similarity of Data to Particles

Some years ago we pointed out that true creation and annihilation in our universe appears in only two arenas:[112] data transformations within computers and particle transformations in Nature. Both subatomic particles and data can be created or destroyed or combined to produce new particles or new data. The similarity is compelling! In the following sections we will amplify these ideas based on the theory of computer languages and computer grammars.

We will also consider particle interactions at a more abstract level and show how to create "sub-atomic particle quantum Turing machines" for both the Standard Model and Superstring theories. (In this appendix we take a neutral stance on whether the Theory of Everything is a Superstring theory or our Complexon Standard Model. The possibility of a deep connection of these two theoretic approaches cannot be ruled out at present.)

---

[112] Blaha (1998). Please note that we will discuss only Nature in this book since theological constructs are beyond the scope of our inquiry.

B.2 Particle Physics Lagrangians define a Language

Physicists use perturbation theory to perform computations in quantum field theories such as The Standard Model. Perturbation theory takes the interaction terms of the Lagrangian and performs approximate calculations of the probabilities of particle interactions. Perturbation theory calculations are normally visualized using Feynman diagrams. For example the collision of two electrons to produce two electrons with different energies and momenta (called electron-electron elastic scattering) can be visualized as a sum of terms corresponding to the various ways the electrons can interact. The number of terms in this example is infinite in principle. Since this infinite sum cannot be calculated the sum is approximated by a finite number of terms. The simplest and, as it turns out, the dominant terms in electron-electron scattering are:

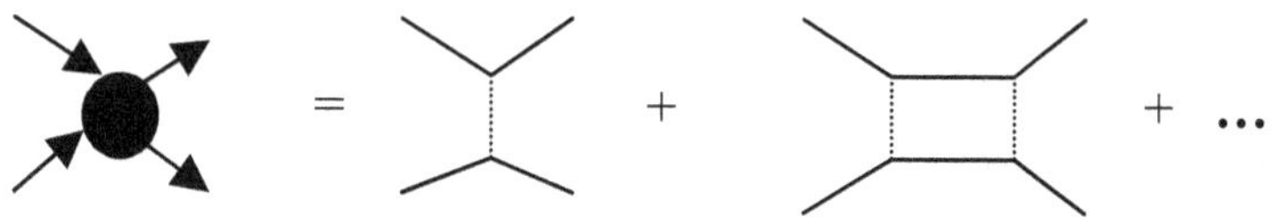

Figure B.1. The first few terms in the approximate perturbation theory calculation of the scattering of two electrons. The dotted lines represent photons that carry the electromagnetic force between the electrons.

We will see that the individual terms (diagrams) in a perturbation theory calculation can be viewed as words. The words are part of a language with an alphabet and grammar.

The fundamental particles of the Standard Model constitute the set of symbols or the alphabet of a language with 52 letters in The Complexon Standard Model. (It is an interesting, but meaningless, coincidence that most alphabet-based human languages have 20 to 40 letters in their alphabets. English has 26 and so on.) As we shall see, the particle language grammar is a quantum extension of a type of computer language (developed by Chomsky and others) that uses *production rules*. The production rules for the grammar of the Standard Model are easily derived from the interaction terms of the Standard Model.

The concept of the language of the Standard Model is very simple. The technical details of the language description require a discussion of computer languages and Quantum Turing Machines.

The following sections describe the basic idea of this linguistic representation of the Standard Model. They show how particle interactions can be viewed as transformations (processes) within a Quantum Computer that accepts the computer language generated by the Standard Model Lagrangian interaction terms.

B.3 A Linguistic Representation of the Standard Model

In this book and earlier works we explored the features of the Standard Model. We saw that it was consistent with almost all the known properties of elementary particles. This section introduces a new view or representation of the Standard model that focuses on its interactions and *shows the Standard Model defines a language similar to a computer language.*

The *alphabet* of the language is the set of elementary particles of the Standard Model. The *words* of the language are quantum states consisting of elementary particles. A bound state of several particles such as a proton (three quarks bound together) is a word. A quantum state consisting of several free particles – particles that are not bound together and that are some distance from each other also constitute a word. In fact the entire universe constitutes one mighty word.[113]

The collision or scattering of particles can be viewed as beginning with a combination of letters corresponding to the set of initial particles – the input string or word. This input string undergoes transformations specified by grammar rules to produce an output string (word) of letters corresponding to the outgoing particles after the collision.

In describing this new view of the Standard Model we will focus on the essentials of the processes of creation, transformation and annihilation of matter ignoring (for the moment) particle spin, momentum, angular momentum and other details that are important in the complete theory of the Standard Model. This approximation may have been valid prior to the Big Bang when the universe might have been a mathematical point. Incorporating particle spin, momentum, angular momentum and so on into a "particle" language is not difficult.

---

[113] Blaha (1998).

The idea of associating physics with computers is not as unconventional as it might appear at first. Feynman[114] viewed computers as relevant for Physics: "If we suppose we know all the physical laws perfectly, of course we don't have to pay any attention to computers. It's interesting anyway to entertain oneself with the idea that we've got something to learn about physical laws; and if I take a relaxed view here ... I'll admit that we don't understand everything." Feynman wanted to simulate physics computations on a quantum computer in the hope that it would be faster than a conventional computer. We will show the Standard Model itself actually defines a specific (theoretical) quantum computer – a far more exciting possibility – because it gives a new view of Reality. Nature itself is a form of computer.

SuperString theory can also be formulated within a Quantum Computer framework.

A computer language representation of particle physics is of great interest in itself. It may generate new insights into the process of matter creation and transformation. It may lead to a new understanding of the fundamental nature of the universe. And it appears to suggest a rationale for approaches such as the currently popular SuperString theories of elementary particles.

B.3.1 Linguistic View of an Interaction

We will begin by looking at the simple interaction term:

$$e\bar{A}e$$

From a computer language perspective this Lagrangian interaction term can be viewed as specifying a set of grammar rules called *production rules*.

In fact each interaction term in the Standard Model Lagrangian can be viewed as specifying a set of grammar rules. The combined set of grammar rules for all Standard Model interaction terms defines a grammar with particles constituting the alphabet (letters or symbols) of the grammar.

To appreciate the mapping (or analogy) between particles and alphabetic letters, and of interaction terms and computer grammar, we have to understand the process of data characters (or letters) flowing through a computer. It is an interesting and little noted fact (because it is viewed as trivial) that a computer

[114] R. P. Feynman, International Journal of Theoretical Physics, **21**, 467 (1982).

can generate (or absorb) data as part of the computation process. For example we might write a computer program that takes a set of letters input into a computer and outputs each input letter twice:

In a sense the computer has created data characters just like particle interactions can create particles. Computers can also absorb (or annihilate) data (usually to our dismay). So we can see an analogy between the transformations of data characters in a computer, and particle annihilation and creation. *Nothing else in Nature is so directly analogous to particle creation and annihilation.*

This observation leads us to take the view that particles are data packets that we denote with letters (symbols). They contain quantum numbers and other properties (mass, spin, momentum, and so on) that certainly are data. And they have a grammar that we summarize with a Lagrangian.

B.3.2 Computer Grammars

The Standard Model Lagrangian in our view specifies a *grammar* in the sense of Naom Chomsky. Chomsky's concept of a language, and of a grammar, has important applications in the theory of computation and computers.

There are four basic types of languages in the Chomsky approach: called type 0, type 1, type 2 and type 3. They differ in the allowed forms of their grammar rules (also called *production rules*). We will be interested in type 0 languages. A type 0 language (also called an *unrestricted rewriting system*) is the most general type of language. It allows any grammar production rule of the form

$$x \longrightarrow y$$

where x and y are strings of characters.

Production rules specify how one string of characters transforms into another string of characters. Calculations in computers using computer languages are reducible to sets of grammar rules for string manipulation that are similar to the one shown above.

Each term in the interaction part of the Standard Model Lagrangian is equivalent to one or more production rules where the characters are particles. *The Standard Model can be viewed as generating a type* 0 *language.* This language goes beyond current types of grammars because it is inherently quantum probabilistic in nature. Quantum aspects of these rules will be described later.

Before looking at the production rules generated by an interaction term in a Lagrangian we will discuss a formal grammar. A grammar is a quadruple of items that is usually symbolized by the expression

<N, T, S, P>

where N is a set of variables called *nonterminal symbols,* T is a set of *terminal symbols,* S is a special nonterminal symbol called the *head* or *start symbol,* and P is a finite set of production rules. The angle braces < and > are merely a mathematician's way of saying these items are grouped together to constitute (or make) a grammar.

The *terminal* symbols are the set of characters that are allowed in input strings or output strings. The *nonterminal* symbols are the set of characters that appear in intermediate steps that lead from the input string to the output string. They are like internal variables or symbols.[115] The combined set of terminal and nonterminal symbols make up the *vocabulary* (alphabet) of a language.

Chomsky's definition of a language is the set of all strings of terminal symbols that can be generated by applying the production rules to the *head* symbol (or *start* symbol) S.[116] The head symbol is the symbol that begins all strings of symbols that can be generated in a language.

A simple example of a language in this approach is a vocabulary or alphabet consisting of the ABC's with words created from these letters according to some set of production rules.

[115115] Their particle analogues are virtual particles that only appear in an evolving interaction but not in the input or output. One could view quark-partons as equivalent to nonterminal particles.

[116] The head symbol is analogous to the vacuum state in quantum field theory. Chomsky's definition specifies the possible vacuum fluctuations that can occur. Each string is a specific vacuum fluctuation. An example is an electron-positron pair momentarily popping out of the vacuum—a vacuum fluctuation.

B.3.3 Generalized Input Chomsky Languages

We will generalize Chomsky's idea of language to be the set of all strings that can be generated from all finite input strings of terminal symbols as well as the *head symbol*. We can also view all particles as generated directly or indirectly at the beginning on the universe. The "Big Bang" (the beginning of the universe) then becomes the primeval head symbol.

We can visualize the application of production rules to transform an input string of terminal characters into an output string of terminal characters as:

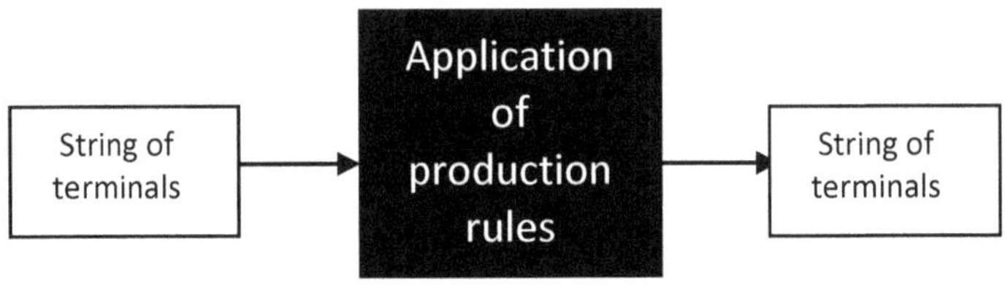

Figure B.2 Generating an output string of terminal characters from an input string of terminal characters using the production rules of a grammar. Inside the black box, transformations of the input string take place and non-terminal symbols may appear and disappear. Non-terminal symbols, by definition, cannot appear in the input or output strings of characters.

In order to make these grammar concepts more concrete we will look at a simple artificial grammar before looking at the grammar generated by interaction terms in the Standard Model. The nonterminal symbols will be the letters S (the head symbol), A and B. The terminal symbols will be the letters x and y. The production rules will be

$S \longrightarrow AB$ Rule I
$A \longrightarrow y$ Rule II
$A \longrightarrow Ay$ Rule III
$B \longrightarrow x$ Rule IV
$B \longrightarrow Bx$ Rule V

The Chomsky computer language that this grammar generates consists of all strings containing any number of y's followed by any number of x's since any of these strings can be generated from the head symbol S using the production rules. Because of rule I the y's are placed to the left and x's are placed to the right. The order of the symbols matters just as it does in human language – consider the words ides and dies which differ only in the order of i and d!

An example of generating a string yyxxx from the head symbol is:

S	$\longrightarrow$ AB	Rule I
AB	$\longrightarrow$ AyB	Rule III
AyB	$\longrightarrow$ yyB	Rule II
yyB	$\longrightarrow$ yyBx	Rule V
yyBx	$\longrightarrow$ yyBxx	Rule V
yyBxx	$\longrightarrow$ yyxxx	Rule IV

The production rule used to make each transition is listed above on the right.

Our generalization of the Chomsky definition of language would allow any string to be the starting point – not just the head symbol S. Using the sample grammar described on the previous page the generalized language becomes any string of x's and y's.

A more interesting language can be created by adding two new rules to the rules on the preceding page:

y	$\longrightarrow$ A	Rule VI
x	$\longrightarrow$ B	Rule VII

The resulting language – the set of strings of terminal symbols – remains the same despite the addition of these new grammar rules. However the number and variety of transitions becomes much larger. For example the following chain of transitions is allowed,

yx $\longrightarrow$ AB $\longrightarrow$ AyBx $\longrightarrow$ AyyBx$\longrightarrow$ AyyyBxx $\longrightarrow$ yyyyxx

In the next section we will extend the concept of computer grammars by allowing probabilistic grammar rules – production rules which have an associated probability of executing.

B.4 Probabilistic Computer Grammars

Grammar, which knows how to control even kings.
Molière - Les Femmes Savantes (1672) Act II, Scene 6

B.4.1 Probabilistic Computer Grammars™

The preceding section described the production rules for a *deterministic grammar*. The left side of each production rule has one, and only one, possible transition.

Non-deterministic grammars allow two or more grammar rules to have the same left side, and different right sides. For example,

$$A \rightarrow y$$
$$A \rightarrow x$$

could both appear in a non-deterministic grammar.

Non-deterministic grammars can be easily (almost "naturally") associated with probabilities. The probabilities can be classical probabilities or quantum probabilities. An example of a simple non-deterministic grammar is specified by the production rules:

S → xy Rule I
x → xx Rule II Relative Probability = .75
x → xy Rule III Relative Probability = .25
y → yy Rule IV

where the head symbol is the letter S, and the terminal symbols are the letters x and y. The relative probability of generating the string xxy vs. the relative probability of generating the string xyy from the string xy is

xy → xxy relative probability = .75
xy → xyy relative probability = .25

The string xxy is three times more likely to be produced than the string xyy.

For each starting string one can obtain the relative probabilities that various possible output strings will be produced.

A more practical example of a Probabilistic Grammar™ can be abstracted from flipping coins – heads or tails occur with equal probability – 50-50. From this observation we can create a little Probabilistic Grammar™ for the case of flipping two coins. Let h represent heads and t represent tails. Then consider the grammar:

S → hh
S → tt
S → ht
S → th

h → t	**probability = .5 (50%)**
h → h	**probability = .5 (50%)**
t → h	**probability = .5 (50%)**
t → t	**probability = .5 (50%)**

The last four rules above embody the statement that flipping a coin yields heads or tails with equal probability (50% or .5).

Now let us consider starting with two heads hh. The possible outcomes and their probabilities are:

hh → hh	**probability = .5 * .5 = .25**
hh → th	**probability = .5 * .5 = .25**
hh → ht	**probability = .5 * .5 = .25**
hh → tt	**probability = .5 * .5 = .25**

If we don't care about the order of the output heads and tails, then the probability of flipping two heads and getting a head and tail (hh → ht or hh → th) is .25 + .25 = .5.

This simple example shows the basic thought process of a non-deterministic grammar with associated probabilities.

The combination of a non-deterministic grammar and an associated set of probabilities for transitions can be called a *Probabilistic Grammar*. We will see that the grammar production rules for the Standard Model must be viewed as constituting a Probabilistic Grammar™ with one difference. The "square roots" of probabilities – probability amplitudes – are specified for the transitions in the grammar. The Standard Model requires probability amplitudes since it is a quantum theory. Therefore we will describe probabilistic grammars with associated probability amplitudes (such as that of the Standard Model) as *Quantum Probabilistic Grammars*.

B.4.2 Quantum Probabilistic Grammar

An example of a Quantum Probabilistic Grammar can be constructed based on an analogy with a two slit photon experiment. Imagine a wall with two slits. A source shoots photons at the wall. A photon can go through either slit with equal quantum probability. An illustration of this experimental arrangement is:

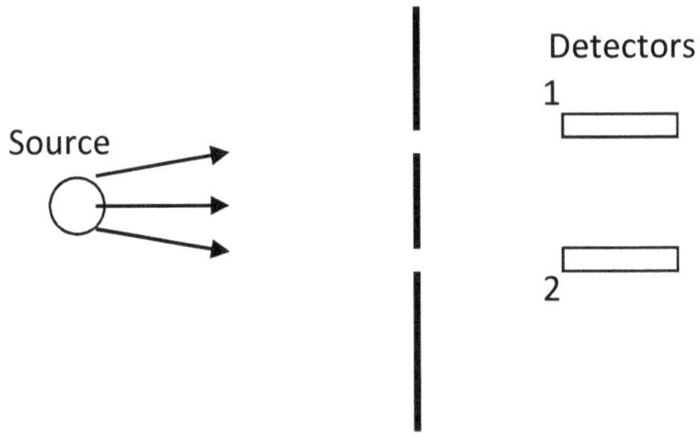

Figure B.3. Two slit photon experimental setup.

A simple Quantum Probabilistic Grammar can be constructed corresponding to this experimental setup:

$$S \rightarrow 1 \quad \text{probability amplitude} = 1/\sqrt{2}$$
$$S \rightarrow 2 \quad \text{probability amplitude} = 1/\sqrt{2}$$

The head symbol S represents the source. The digit 1 represents a photon going through slit 1. The digit 2 represents a photon going through slit 2.

The values of the probability amplitudes $1/\sqrt{2}$ can be calculated using Quantum Mechanics. The probability for a photon to go through slit 1 is the absolute value squared of the probability amplitude:

$$\text{Probability to go through slit 1} = (1/\sqrt{2})^2 = .5$$

and the probability for a photon to go through slit 2 is

$$\text{Probability to go through slit 2} = (1/\sqrt{2})^2 = .5$$

This simple example illustrates the basics of a Quantum Probabilistic Grammar.

Before applying these concepts to the Standard Model we will look at a simpler Quantum Field Theory called a ϕ^3 ("phi cubed") theory (ϕ is the Greek letter phi). This theory describes a self-interacting spin 0 particle with no internal symmetries. This theory is a stepping stone to the far more complex Standard Model Quantum Field Theory. We are only interested in it as a simple example of quantum probabilistic grammar rules.

The ϕ^3 theory is so named because it has a cubic Lagrangian interaction term. (Note the exponent 3.) The grammar rules for the ϕ^3 theory are:

$$\phi \ \rightarrow \phi\phi \qquad \text{Rule I}$$
$$\phi\phi \rightarrow \phi \qquad \text{Rule II}$$

Rule I corresponds to the emission of a ϕ particle and rule II corresponds to the absorption of a ϕ particle. We will not introduce a start symbol. Instead we will consider the transitions from an input state of a number of ϕ particles to an output state of (possibly) a different number of ϕ particles. *We will ignore the momenta of the particles.* (This assumption is equivalent to assuming the ϕ particles have infinite mass.)

We will assume either transition above takes place with a "relative probability amplitude" g. We will call this simplified theory the *modified* ϕ^3 *theory*. We will view g as a measure of the probability amplitude for an absorption or emission of a ϕ particle. (g is similar to a coupling constant in Quantum Field Theory.) The probabilities have to be normalized or rescaled so that the sum of all probabilities equals one.

To get a feel for the Quantum Probabilistic Grammar approach we will look at the case of an input state consisting of two ϕ particles. The output states can have one ϕ particle, two ϕ's, three ϕ's, and so on. Each possible output state has a certain probability of occurring. The sum of the probabilities for producing all possible output states must equal one. (Remember that the sum of all possible outcomes of flipping a coin is one. Having it come up heads has probability ½ and having it come up as a tails has probability ½ also.)

The simplest string transition from a two ϕ "input" state to a one ϕ "output" state is:

$$\phi\phi \rightarrow \phi$$

using Rule II. The probability amplitude of this transition is g by assumption.

The transitions between strings can be visualized with diagrams that are like the Feynman diagrams that used in Quantum Field Theory perturbation theory calculations. These diagrams are not the same as Feynman diagrams because they embody time orderings of emissions and absorptions of ϕ particles. (They actually harken back to the time-ordered diagrams that were used by physicists prior to 1950.) **This feature supports the need for the Principle of Asynchronicity described in appendix A.**

In some simple cases the time ordering is irrelevant. For example, the Feynman-like diagram for the simplest case of a two ϕ input state transitioning *directly* to a one ϕ output state is the same as the Feynman diagram:

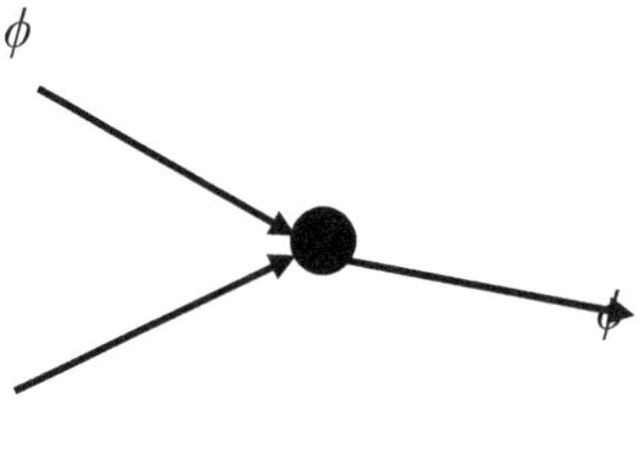

Figure B.4. Diagram for $\phi\phi \rightarrow \phi$. The input states are always on the left and the output states are always on the right in Feynman and Feynman-like diagrams.

The time order of emission and absorption of ϕ particles can be symbolized using parentheses. For example,

$$(\phi)\phi \rightarrow (\phi\phi)\phi = \phi(\phi\phi) \rightarrow \phi(\phi) = \phi\phi \qquad \text{Diagram A (Fig. B.5)}$$

and

$\phi(\phi) \rightarrow \phi(\phi\phi) = (\phi\phi)\phi \rightarrow (\phi)\phi = \phi\phi$ Diagram B (Fig. B.5)

These string transitions correspond to different time-ordered Feynman-like diagrams:

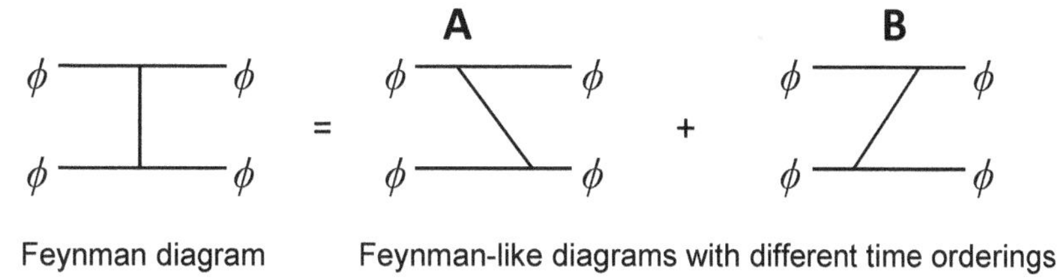

Figure B.5 The true Feynman diagram is the "sum" of the two time-ordered Feynman-like diagrams.

The correspondence between Feynman-like diagrams and the transitions between strings based on the Quantum Probabilistic Grammar can be seen by taking vertical slices on diagrams A or B above after each emission or absorption. For example,

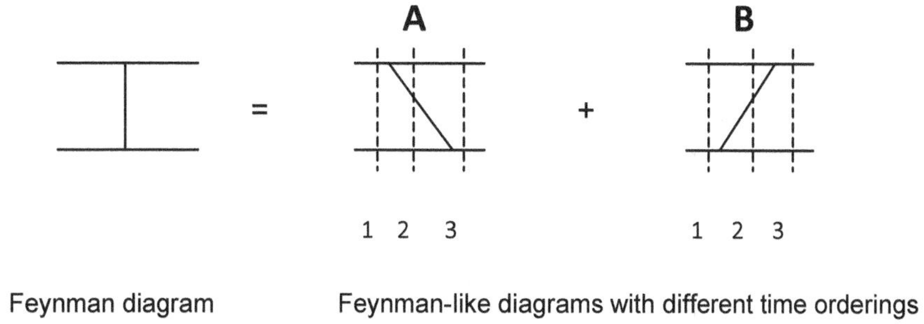

Figure B.6 Slicing Feynman-like diagrams. A slice is made after each emission or absorption. As you read down a slice the particles are listed in the same order as the corresponding string. Each slice is numbered starting from the left.

The string corresponding to each numbered slice in the above figures is similarly numbered in the following transitions:

Slice:	1		2		3	
	$(\phi)\phi$	$\rightarrow$	$(\phi\phi)\phi = \phi(\phi\phi)$	$\rightarrow$	$\phi(\phi) = \phi\phi$	**A**
Slice:	1		2		3	
	$\phi(\phi)$	$\rightarrow$	$\phi(\phi\phi) = (\phi\phi)\phi$	$\rightarrow$	$(\phi)\phi = \phi\phi$	**B**

Parentheses on the left side of an arrow indicate the particle(s) that emits a new particle(s) appearing within the corresponding parentheses on the right side of the arrow.

A transition from an input state containing ϕ particles to an output state containing ϕ particles always has an infinite number of ways of taking place and thus an infinite number of Feynman-like diagrams. Readers familiar with the perturbation theory of Quantum Field Theory will remember that these diagrams are the same as the Feynman diagrams generated by perturbation theory with the additional feature of having time orderings.

We will now look at the transition of two ϕ particles to two ϕ particles: $\phi\phi \rightarrow \phi\phi$. There are an infinite number of Feynman-like diagrams for this transition. Some of the simpler Feynman diagrams are:

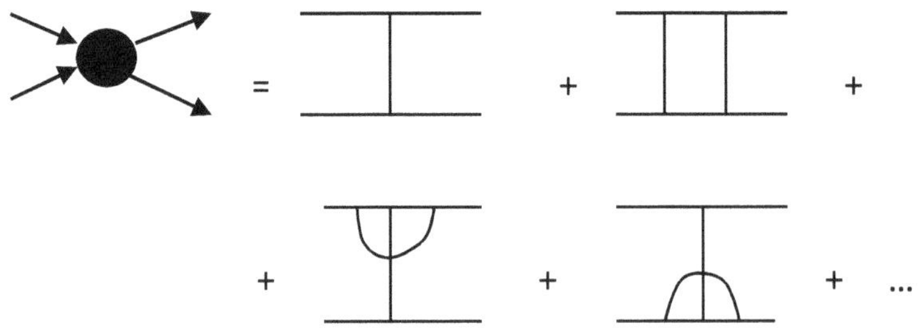

Figure B.7. Diagrams for an elastic collision with two incoming particles and two outgoing particles.

Each Feynman diagrams corresponds to several time-ordered Feynman-like diagrams.

In evaluating these diagrams to calculate probabilities we must remember that we are ignoring space-time aspects such as particle propagators and momenta. So the calculation of the probability amplitude for this process becomes a counting problem of the number of diagrams that exist for each power of g^2. The probability amplitude for each diagram is a power of g^2.

Counting diagrams is a combinatorial mathematics problem that we will not explore in detail because it is peripheral to our interests. Consequently we will simply express the probability amplitude as:

$$A_2(g) = \sum_{n=1}^{\infty} a_n g^{2n}$$

where the mathematical expression on the right represents a sum from n = 1 to infinity and where the numbers a_n are integer numbers equal to the number of different diagrams having a power of g^{2n} as its probability amplitude. Each intersection of lines (called a vertex) contributes a factor of g to the amplitude for that diagram. The powers of g for the simpler diagrams that appear on the previous page are:

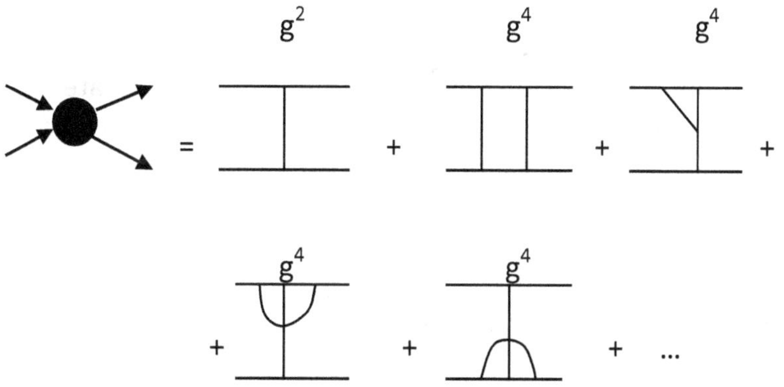

Figure B.8. The power of g for some simple diagrams.

The value of the first constant a_1 is 1 since there is only one Feynman diagram with amplitude value g^2 for this process – the first diagram to the right of the = sign in Fig. B.8. We treat all time-ordered variations of a Feynman diagram as contributing one to the value of a_n. The value of a_n grows rapidly as n increases. For large n the value of a_n is of the order of $(n!)^2$. Consequently the sum is an asymptotic power series. Here again the details are not important for us. We are not looking for a numerical result.

The (unnormalized) relative probability for the transition $\phi\phi \rightarrow \phi\phi$ is:

$$P_2 = |A_2(g)|^2$$

where |...| represents the absolute value of the complex probability amplitude. In quantum theories a probability is always the square of the absolute value of its probability amplitude. The probability P_2 is a relative probability that must be normalized – multiplied by a factor that makes the sum of the probabilities of all possible outcome states equal to one. To calculate this probability we must calculate the sum of all the relative probabilities P_n to produce any number of ϕ particles from a two ϕ input state.

$$\phi\phi \rightarrow \phi \ldots$$

The total of the relative probabilities is:

$$P = \sum_{n=1}^{\infty} P_n$$

where P_n is the relative probability to produce an output state with n ϕ particles.

The calculation of the relative probabilities P_n for n ϕ particles output states is similar to the calculation P_2. For example, for three particles

$$A_3(g) = \sum_{n=1}^{\infty} b_n \, g^{2n+1}$$

where the numbers b_n count the number of distinct diagrams with the power g^{2n+1} and

$$P_3 = |A_3(g)|^2$$

The absolute (normalized) probability to produce an n ϕ particle output state is

$$Q_n = P_n/P$$

The sum of all possible output state probabilities equals one:

$$1 = \sum_{n=1}^{\infty} Q_n$$

The modified ϕ^3 Quantum Field Theory provides a simple example of a Quantum Probabilistic Grammar. We will now turn to the Standard Model and examine its Quantum Probabilistic Grammar. Because it encompasses a much larger number of different particles (letters) and interactions (grammar rules) it will be significantly more complex.

B.4.3 Probability Amplitudes of Quantum Grammars

Quantum Grammar rules associate a probability amplitude with each production rule. To find the probability for a transition from an initial state to a final state we must calculate the probability amplitude for the transition through the repeated application of the production rules for each possible path from the initial state to the final state. We assume each initial state of the Quantum Turing machine begins with probability amplitude one. (This is a normalization condition for the initial state in reality.)

When a production rule is applied to a state to produce a transition to a new state the current probability amplitude is multiplied by the probability amplitude of the production rule. Thus the total relative probability for the passage from an initial state i to a specific final state f is

$$P_{f\,i} = |\sum_{paths} a_1\, a_2\, a_3\ \ldots\ a_{n(path)}|^2$$

where the sum is over all finite paths that lead from the initial state to the final state through the application of all relevant production rules, and where $a_1 a_2\, a_3 \ldots a_{n(path)}$ is the product of the probability amplitudes of the production rules for each individual path. P_{f_i} is similar in form to a Feynman path integral expression (See Feynman (1965)).

The value of n is path dependent and thus denoted n(path). The relative probability is the absolute value squared of the sum of products of the amplitudes. The relative probability must be normalized to produce an absolute probability such that the sum of the absolute probabilities of all possible final states is one:

$$P_{absolute\text{-}f\,i} = NP_{f\,i}$$

$$N = \sum_f P_{f\,i}$$

$$1 = \sum_f P_{absolute\text{-}f\,i}$$

where the sums are over all possible final states f.

Thus we have a well-defined method for calculating the probability of a transition from an initial state to a final state that is illustrated by the preceding examples. The fact that it bears some resemblance to the path integral methods for quantum mechanics pioneered by Feynman suggests that Quantum Grammars are of interest to physics. This author was first struck by the similarity of string transitions via production rules to path integrals in 1981. After all, a jagged (discrete) Feynman path is really a string of coordinates marking the end points of each line segment of which the path is composed. Thus each path in a Feynman sum over paths can be represented by a string. The evolution of a path from line segment to line segment can be viewed as the repeated application of a probabilistic production rule. The path sum equivalent of the probability amplitude of a production rule is an exponential Hamiltonian factor that is a function of the change in string coordinates "due to the production rule."

B.5 Standard Model Quantum Grammar

B.5.1 Grammar Production Rules of Quantum Electrodynamics

We now consider the grammar production rules of the Quantum Electrodynamics (electromagnetism) sector of the Standard Model. The production rules corresponding to the electromagnetic interaction term for electrons and positrons in the Standard Model

$$\bar{e}Ae$$

are:

Electron-Positron QED Production Rules

e → eA
e → Ae
eA → e
Ae → e
p → pA
p → Ap
Ap → p
pA → p
ep → A

$$\text{pe} \rightarrow \text{A}$$
$$\text{A} \rightarrow \text{ep}$$
$$\text{A} \rightarrow \text{pe}$$

where e represents an electron, p represents a positron, and A represents a photon. The production rules describe the emission and absorption of photons by electrons and positrons as well as the annihilation of an electron and positron to produce a photon, and the decay of a photon into an electron-positron pair.

An example of an interaction between two electrons in the linguistic approach is:

$$\begin{array}{ccccc} 1 & & 2 & & 3 \\ \text{ee} & \rightarrow & \text{eAe} & \rightarrow & \text{ee} \end{array}$$

where the electrons interact by exchanging one photon. One Feynman-like diagram for these transitions is:

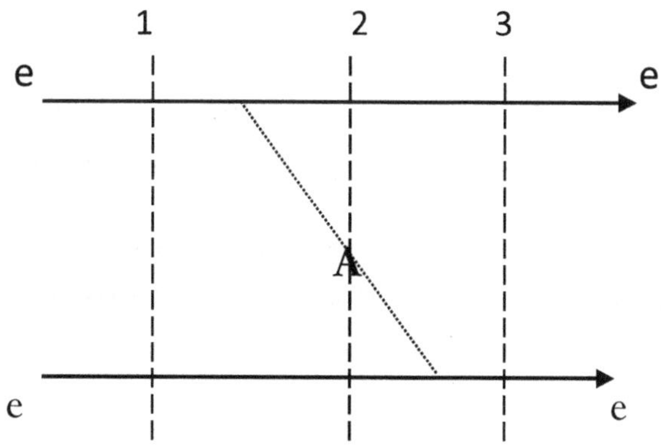

Figure B.9. A diagram showing how two electrons interact by exchanging a photon. As time increases the electrons move from left to right. The upper electron emits the photon. This corresponds to the left e transitioning to eA using the grammar rule e $\rightarrow$ eA. (There is a similar diagram Fig. B.10 in which the lower electron emits the photon.)

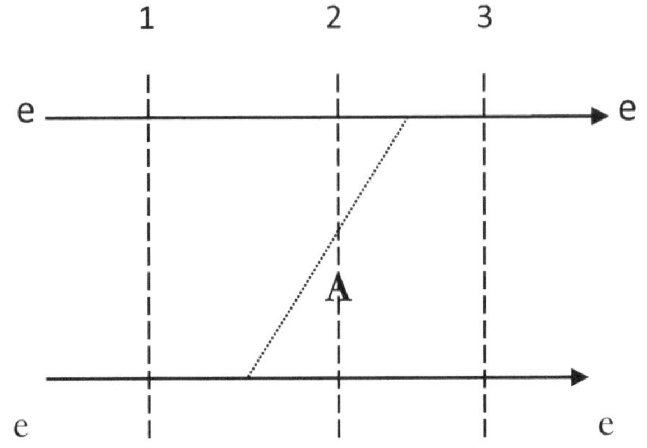

Figure B.10. Another diagram with two electrons interacting by exchanging a photon. In this case the lower electron emits the photon. This corresponds to the right e in the initial ee string transitioning to Ae using the grammar rule e $\rightarrow$ Ae.

The vertical slices in the Feynman diagram which are numbered 1, 2 and 3 correspond to the three numbered strings in the transitions generated from the production rules. Each string has an ordering that corresponds to the order of particles as you descend a slice. For example slice 2 in Fig. B.10 has an electron, photon, and another electron in that order as you descend corresponding to string 2 above.

Another Feynman-like diagram that contributes to this process has the lower electron emitting a photon that is then absorbed by the upper electron. The Feynman diagram for this process represents the sum of both of the previous diagrams:

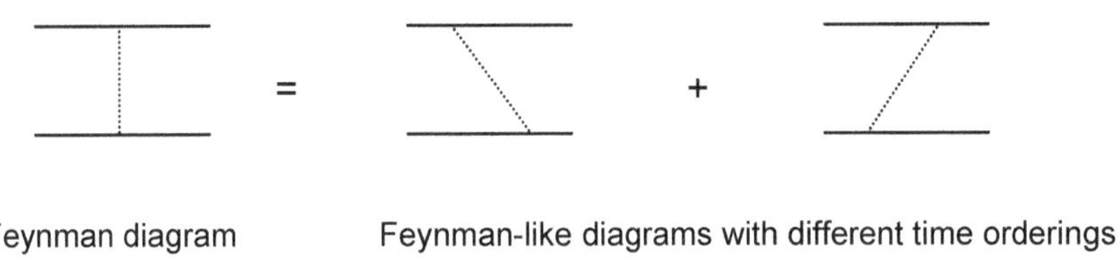

Figure B.11. A Feynman diagram represents several of our Feynman-like diagrams with different time orderings of particle emission and absorption.

A more complex example of a Feynman-like diagram appears in Fig. B.12. Six slices appear corresponding to the various intermediate states in this complex electron-electron interaction. The production rules can be used to generate a sequence of strings that correspond to the slices:

$$\begin{array}{cccccc} 1 & 2 & 3 & 4 & 5 & 6 \\ \text{ee} \rightarrow & \text{eAAe} \rightarrow & \text{eAe} \rightarrow & \text{eAAe} \rightarrow & \text{eepAe} \rightarrow & \text{eepe} \end{array}$$

As you descend each slice the particles are ordered in the same way as the corresponding string. For example, as you descend slice 5 the order of the particles is electron, electron, positron, photon and electron, and the string is eepAe.

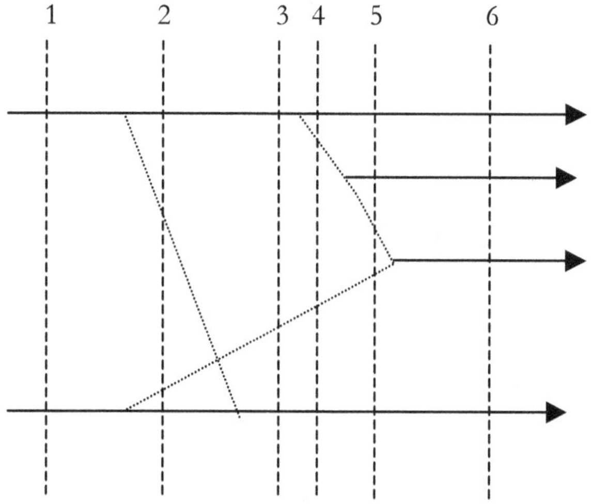

Figure B.12. A diagram for the collision of two electrons that produce a new electron-positron pair: ee → eepe.

The transitions between character strings have an ambiguity. For example in the above transition

$$\begin{array}{cc} \text{eAAe} \rightarrow & \text{eAe} \\ 2 & 3 \end{array}$$

could have taken place through (eA)Ae $\rightarrow$ (e)Ae with the left (upper) e absorbing an A or through eA(Ae) $\rightarrow$ eA(e) with the right (lower) e absorbing an A. (Parentheses are used for grouping to show which electron absorbed the photon.) This ambiguity reflects the fact that there are several possible time orderings. The preceding diagram actually corresponds to eA(Ae) $\rightarrow$ eA(e). The right (lower) electron absorbs the photon.

To calculate the probability of an actual physical transition occurring such as ee $\rightarrow$ eepe we must take account of all diagrams with all possible time orderings for the specified input and output states. This is a monumental chore since an infinite number of diagrams are involved. Normally only the simplest diagrams are evaluated since they dominate the electromagnetic interactions of electrons and positrons.

The above examples show how the electromagnetic interaction part of the Standard Model lagrangian can be viewed as defining a grammar. The grammar has corresponding Feynman-like diagrams. As we pointed out earlier, these types of diagrams have time orderings that are similar to the time orderings that appeared in perturbation theory calculations before 1950.

The Weak Interaction and Strong Interaction parts of the Standard Models also define grammars. Consequently we can view the complete Standard Model Lagrangian as defining a grammar where the "letters" (alphabet or vocabulary) are the elementary particles of the model and the Feynman-like diagrams corresponding to the Standard Model are a sequence of strings generated by applying the production rules specified by the Standard Model Lagrangian.

B.5.2 Production Rules for the Weak and Strong Interactions

The Weak and Strong interaction terms in the Standard Model Lagrangian are also easily translated into grammar production rules (although the process is laborious since there are so many of them). We will illustrate these cases using the Weak interaction terms:

$$\bar{\nu}_e W^- e$$

and

$$\bar{\nu}_\mu W^- \mu$$

where ν_e represents an electron neutrino, ν_μ represents a muon neutrino, μ represents a muon, W^- is a gauge field of the Weak interaction and e is an electron; and the Strong interaction term

$$\bar{u}Gu$$

where u is a u quark and G represents gauge fields of the Strong interaction.

Notice that there are several types of neutrinos: electron neutrinos, muon neutrinos and tau neutrinos. The three kinds of neutrinos have different internal quantum numbers that distinguish them. Neutrinos do not have electromagnetic charge. They are neutral as their name suggests. Each kind of neutrino has a corresponding charged partner. We are familiar with the electron. The other charged partners are the muon and tau particle. These charged particles are like heavy electrons for the most part. The three charged leptons also have distinguishing internal quantum numbers.

The preceding interaction terms imply production rules such as:

$$e \rightarrow W^- \nu_e$$

$$e \rightarrow \nu_e W^-$$

$$W^- \rightarrow e\, \nu_e$$

$$W^- \rightarrow \nu_e\, e$$

$$\mu \rightarrow \nu_\mu W^-$$

$$\mu \rightarrow W^- \nu_\mu$$

$$u \rightarrow Gu'$$

$$u \rightarrow u'G$$

and so on where e is an electron, p is a positron, W^- is a negative W gauge boson, ν is a neutrino, G is a Strong interaction gauge boson and u and u' are u quarks which may have different color quantum numbers.

These production rules generate string equivalents of Weak interaction transitions such as muon decay:

$$\begin{array}{ccccc} 1 & & 2 & & 3 \\ \mu & \rightarrow & W^- \nu_\mu & \rightarrow & e\nu_e\nu_\mu \end{array}$$

which has the corresponding Feynman-like diagram:

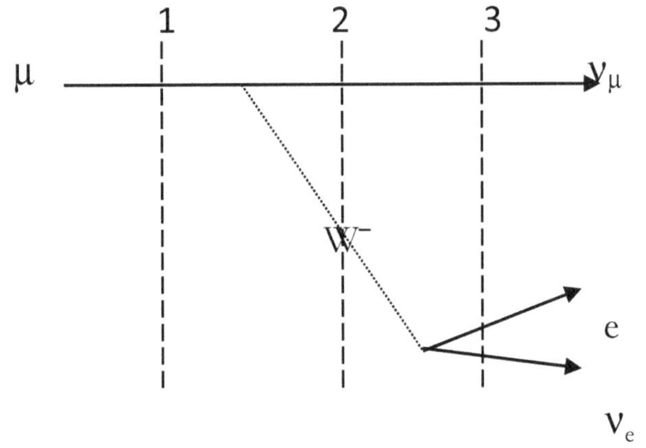

They also generate string equivalents of Strong interaction transitions between quarks and color gauge fields such as:

$$\begin{array}{ccccc} 1 & & 2 & & 3 \\ uu & \rightarrow & uGu & \rightarrow & u'u' \end{array}$$

with the corresponding Feynman-like diagram:

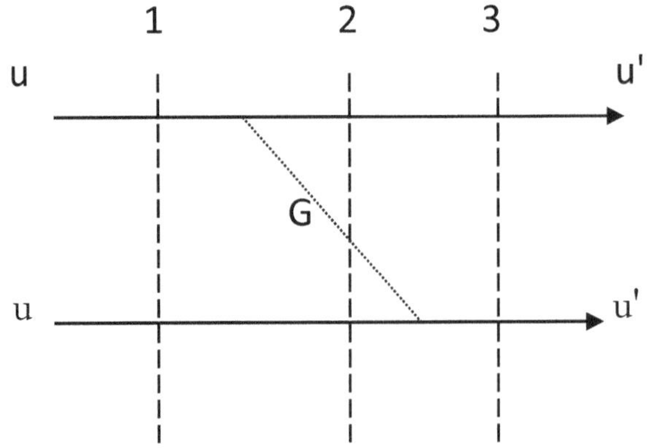

The preceding examples are among the simplest cases of the infinite variety of Feynman-like diagrams that can be generated from the Standard Model production rules.

B.5.3 The Standard Model Quantum Grammar

At this point we have created a view of the Standard Model of elementary particles in which the particles are an alphabet of perhaps 52 letters. The particle alphabet can be combined into strings that represent input or output states of scattering particles as well as bound states. Transitions between strings take place through quantum grammar production rules and correspond to time-ordered Feynman-like diagrams.

So we now have a map between the Standard Model and a language, with letters, words and a grammar; and an interpretation of the language in terms of rules of calculation and experimental setups.

The linguistic representation of the Standard Model that we have developed omits many important calculation details as well as important particle properties such as spin and particle momenta. These features could be added in a direct way. The focus of our investigation is on the essentials of the acts of creation and annihilation of particles in particle interactions.

The character string transition approach based on production rules is equivalent to the Feynman diagram approach. It does however provide a different, and simpler, view. Interestingly the Feynman diagram approach to calculations is very difficult and tedious – but it often leads to a simple result due to the massive cancellation of many complex terms with each other. (An attempt to simplify perturbation theory diagram calculations was made by Cheng and Wu in the early 1970's and by this author privately.) The simple linguistic approach might be a hint of a more efficient way of calculating in quantum field theories like the Standard Model where complications are absent from the very beginning.

Whether or not the linguistic approach leads to a less complicated theory or method of calculation remains to be seen. However we have now obtained a rather amazing result. After 2500 years of speculation on the nature of matter we have developed a surprisingly simple theory (for everything except gravitation) called the Standard Model that can be viewed, in part, as a quantum type 0 computer language. It has an alphabet (vocabulary) of roughly 52 particles, and a set of production rules specified by the interaction part of the Standard Model Lagrangian.

This situation represents something of a miracle. There is no reason that Nature should have so few particles that interact with each other through a simple set of rules. A computer language theorist would call the language of the Standard Model a language with a finite representation. Simply put, this means the words of the language can be generated from a finite vocabulary (alphabet or set of particles) and a finite set of production rules.

B.5.4 The Standard Model Language is Surprisingly Simple

Many physicists have felt that there are too many elementary particles in the Standard Model. From the point of view of a language theorist *the finite language representation of the Standard Model is a very special situation.* As Hopcroft and Ullman[117] point out, "there are many more languages than finite representations." Languages can have infinite alphabets or infinite sets of production rules or other complications.

The physical equivalent of an infinite alphabet would be a universe with an infinite number of different types of matter. Every particle of matter could have a different mass and differ in other properties. From this point of view the Standard Model is truly a marvel of simplicity.

The Standard Model, in fact, is a very compact, finite description of most of the known features of our universe. The linguistic view of the Standard Model suggests we should view elementary particles as symbols or clumps of data – a vocabulary. The interactions of the elementary particles involve the creation or annihilation of particles – creation in the deepest form seen by man.

Remarkably, the data flowing through a computer can be viewed as being transformed from one form to another and being output to different destinations. Data flowing through a computer can be divided into streams that can be sent to different output channels such as a printer or the computer screen. For example, the character, 'a', can be a data item in a stream of data that is sent to both a printer and the screen:

---

[117] J. E. Hopcroft and J. D. Ullman, *Formal Languages and Their Relation to Automata*, (Addison-Wesley, Reading, MA, 1975) page 2.

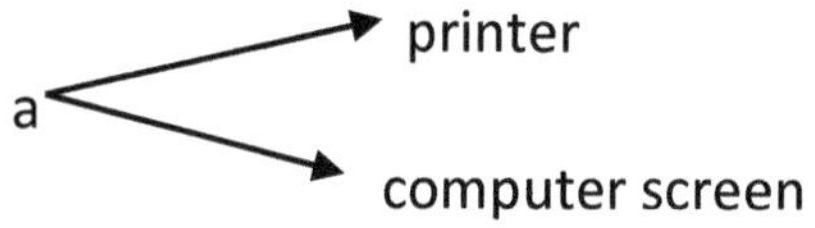

The streaming of data characters to different output channels is quite analogous to the output of particles from particle scattering as we have seen.

The simple production rules that describe particle scattering in the Standard Model suggest a fundamental simplicity at the core of Reality that goes far beyond the speculations of philosophers and scientists of earlier ages.

Appendix C. Physical Constants Used in Calculating Numerical Expressions

Some physical constants that we found to be of use in the evaluation of expressions are (assuming units with $c = \hbar = 1$):

$$G \equiv 7.39 \times 10^{-29}\ \text{gm}^{-1}\ \text{cm} \tag{C.1}$$
$$G \equiv 2.59 \times 10^{-66}\ \text{cm}^{2} \tag{C.2}$$
$$G \equiv 2.91 \times 10^{-87}\ \text{s}^{2} \tag{C.3}$$

$$H_0 \equiv 2.133 \times 10^{-33} h\ \text{ev} \tag{C.4}$$
$$H_0 \equiv 1.08 \times 10^{-28} h\ \text{cm}^{-1} \tag{C.5}$$
$$H_0 \equiv 3.24 \times 10^{-18} h\ \text{s}^{-1} \tag{C.6}$$
$$h\ = .66 \tag{C.7}$$

$$GH_0 \equiv 7.98 \times 10^{-57} h\ \text{gm}^{-1} \tag{C.8}$$

$$\rho_{cr} \equiv 1.88 \times 10^{-29} h^2\ \text{gm cm}^{-3} \tag{C.9}$$

$$M_{Planck} \equiv 1.22 \times 10^{28}\ \text{ev} \tag{C.10}$$
$$M_{Planck} \equiv 2.18 \times 10^{-5}\ \text{g} \tag{C.11}$$

$$M_{Planck} \equiv 6.20 \times 10^{32}\ \text{cm}^{-1} \tag{C.12}$$
$$\text{Planck Length} = M_{Planck}^{-1} \equiv 1.61 \times 10^{-33}\ \text{cm} \tag{C.13}$$

$$M_{Planck} \equiv 1.85 \times 10^{43}\ \text{s}^{-1} \tag{C.14}$$
$$\text{Planck time} = M_{Planck}^{-1} \equiv 5.41 \times 10^{-44}\ \text{s} \tag{C.15}$$

$$1\ \text{ev} \equiv 5.08 \times 10^{4}\ \text{cm}^{-1} \tag{C.16}$$
$$1\ \text{ev} \equiv 1.52 \times 10^{15}\ \text{s}^{-1} \tag{C.17}$$
$$1\ \text{ev} \equiv 1.79 \times 10^{-33}\ \text{g} \tag{C.18}$$

$$1\text{ g} \equiv 2.85 \times 10^{37}\text{ cm}^{-1} \tag{C.19}$$

$$\kappa \equiv 4.38\ {}^{\circ}\text{K}^{-1}\text{ cm}^{-1} \tag{C.20}$$
$$\kappa \equiv 1.31 \times 10^{11}\ {}^{\circ}\text{K}^{-1}\text{ s}^{-1} \tag{C.21}$$
$$\kappa \equiv 8.62 \times 10^{-5}\text{ ev }{}^{\circ}\text{K}^{-1} \tag{C.22}$$

$$1\text{ Gyr} = 3.16 \times 10^{16}\text{ s} \tag{C.23}$$

where κ is Boltzmann's constant.

REFERENCES

Blaha, S., 1998, *Cosmos and Consciousness* (Pingree-Hill Publishing, Auburn, NH, 1998).

_______2004, *Quantum Big Bang Cosmology: Complex Space-time General Relativity, Quantum Coordinates™ Dodecahedral Universe, Inflation, and New Spin 0, ½, 1 & 2 Tachyons & Imagyons* (Pingree-Hill Publishing, Auburn, NH, 2004).

______ *2005a, Quantum Theory of the Third Kind: A New Type of Divergence-free Quantum Field Theory Supporting a Unified Standard Model of Elementary Particles and Quantum Gravity based on a New Method in the Calculus of Variations* (Pingree-Hill Publishing, Auburn, NH, 2005).

______, 2005b, *The Metatheory of Physics Theories, and the Theory of Everything as a Quantum Computer Language* (Pingree-Hill Publishing, Auburn, NH, 2005).

______, 2005c, *The Equivalence of Elementary Particle Theories and Computer Languages: Quantum Computers, Turing Machines, Standard Model, Superstring Theory, and a Proof that Gödel's Theorem Implies Nature Must Be Quantum* (Pingree-Hill Publishing, Auburn, NH, 2005).

______, 2006, *A Derivation of ElectroWeak Theory based on an Extension of Special Relativity; Black Hole Tachyons; & Tachyons of Any Spin.* (Pingree-Hill Publishing, Auburn, NH, 2006).

______, 2007b, *The Origin of the Standard Model: The Genesis of Four Quark and Lepton Species, Parity Violation, the ElectroWeak Sector, Color SU(3), Three Visible Generations of Fermions, and One Generation of Dark Matter with Dark Energy* (Pingree-Hill Publishing, Auburn, NH, 2007).

______, *2008a, A Direct Derivation of the Form of the Standard Model From GL(16) (Pingree-Hill Publishing, Auburn, NH, 2008).*

______, 2008b, *A Complete Derivation of the Form of the Standard Model With a New Method to Generate Particle Masses Second Edition* (Pingree-Hill Publishing, Auburn, NH, 2008)

______, 2009, *The Algebra of Thought & Reality: The Mathematical Basis for Plato's Theory of Ideas, and Reality Extended to Include A Priori Observers and Space-Time Second Edition* (Pingree-Hill Publishing, Auburn, NH, 2009).

______, 2010a, *Operator Metaphysics: A New Metaphysics Based on a New Operator Logic and a New Quantum Operator Logic that Lead to a Mathematical Basis for Plato's Theory of Ideas and Reality* (Pingree-Hill Publishing, Auburn, NH, 2010).
______, 2010b, *The Standard Model's Form Derived from Operator Logic, Superluminal Transformations and GL(16)* (Pingree-Hill Publishing, Auburn, NH, 2010).
______, 2011a, *21st Century Natural Philosophy Of Ultimate Physical Reality* (McMann-Fisher Publishing, Auburn, NH, 2011).
______, 2011b, *All the Universe! Faster Than Light Tachyon Quark Starships & Particle Accelerators with the LHC as a Prototype Starship Drive Scientific Edition* (Pingree-Hill Publishing, Auburn, NH, 2011).
______, 2011c, *From Asynchronous Logic to The Standard Model to Superflight to the Stars* (Blaha Research, Auburn, NH, 2011).
______, 2012a, *From Asynchronous Logic to The Standard Model to Superflight to the Stars volume 2: Superluminal CP and CPT, U(4) Complex General Relativity and The Standard Model, Complex Vierbein General Relativity, Kinetic Theory, Thermodynamics* (Blaha Research, Auburn, NH, 2012).
______, 2012b, *Standard Model Symmetries, And Four And Sixteen Dimension Complex Relativity; The Origin Of Higgs Mass Terms* (Blaha Reasearch, Auburn, NH, 2012).
Eddington, A. S., 1952, *The Mathematical Theory of Relativity* (Cambridge University Press, Cambridge, U.K., 1952).
Fant, Karl M., 2005, *Logically Determined Design: Clockless System Design With NULL Convention Logic* (John Wiley and Sons, Hoboken, NJ, 2005).
Gradshteyn, I. S. and Ryzhik, I. M., 1965, *Table of Integrals, Series, and Products* (Academic Press, New York, 1965).
Sakurai, J. J., 1964, *Invariance Principles and Elementary Particles* (Princeton University Press, Princeton, NJ, 1964).
Weinberg, S., 1995, *The Quantum Theory of Fields Volume I* (Cambridge University Press, New York, 1995).
Weyl, H., 1950, *Space, Time, Matter* (Dover, New York, 1950).
Weyl, H., (Tr. S. Pollard et al), 1987, *The Continuum* (Dover Publications, New York, 1987).

INDEX

About the Author

Stephen Blaha is an internationally known physicist with interests in Science, the Arts, and Technology. He had an Alfred P. Sloan Foundation scholarship in college. He received his Ph.D. in Physics from Rockefeller University. He has served on the faculties of several major universities. He was also a Member of the Technical Staff at Bell Laboratories, a manager at the Boston Globe Newspaper, a Director at Wang Laboratories, and President of Blaha Software Inc and of Janus Associates Inc. (NH).

Among other achievements he was a co-discoverer of the "r potential" for heavy quark binding developing the first (and still the only demonstrable) non-abelian gauge theory with an "r" potential; first suggested the existence of topological structures in superfluid He-3; first proposed Yang-Mills theories would appear in condensed matter phenomena with non-scalar order parameters; first developed a grammar-based formalism for quantum computers and applied it to elementary particle theories; first developed a new form of quantum field theory without divergences (thus solving a major 60 year old problem that enabled a unified theory of the Standard Model and Quantum Gravity without divergences to be developed); first developed a formulation of complex General Relativity based on analytic continuation from real space-time; first developed a generalized non-homogeneous Robertson-Walker metric that enabled a quantum theory of the Big Bang to be developed without singularities at $t = 0$; first generalized Cauchy's theorem and Gauss' theorem to complex, curved multi-dimensional spaces; received Honorable Mention in the Gravity Research Foundation Essay Competition in 1978; first developed a physically acceptable theory of faster-than-light particles; first showed a universe with three complex spatial dimensions is icosahedral; first derived a composition of extrema method in the Calculus of Variations; first quantitatively suggested that inflationary periods in the history of the universe were not needed; first proved Gödel's Theorem implies Nature must be quantum; provided a new alternative to the Higgs Mechanism, and Higgs particles, to generate masses; first showed how to resolve logical paradoxes including Gödel's Undecidability Theorem by developing Operator Logic and Quantum Operator Logic; first developed a quantitative harmonic oscillator-like model of the life cycle, and interactions, of civilizations; first showed how equations describing superorganisms also apply to civilizations; and first developed an axiomatic derivation of the forms of The Standard Model with DARK PARTICLEs from geometry – space-time properties – The faster than light Standard Model.

He has had a major impact on a succession of elementary particle theories: his Ph.D. thesis (1970), and papers, showed that quantum field theory calculations to all orders in ladder approximations could not give scaling deep inelastic electron-nucleon scattering. He later showed the eigenvalue equation for the fine structure constant α in Johnson-Baker-Willey QED had a zero at $\alpha = 1$ not 1/137 by solving the Schwinger-Dyson equations to all orders in an approximation that agreed with exact results to 8th order in α thus ending interest in this theory. In 1979 at Prof. Ken Johnson's (MIT) suggestion he calculated the proton-neutron mass difference in the MIT bag model and found the result had the wrong sign reducing interest in the bag model. These results all appear in Physical Review papers. In the 2000's he repeatedly pointed out the shortcomings of SuperString theory and showed that The Standard Model's form could be derived from space-time geometry by an extension of Lorentz transformations to faster than light transformations. This deeper space-time basis greatly increases the possibility that it is part of THE fundamental theory.

In the early 1980's Blaha was also a pioneer in the development of UNIX for financial, scientific and Internet applications: benchmarked UNIX versions showing that block size was critical for UNIX performance, developing financial modeling software, starting database benchmarking comparison studies, developing Internet-like UNIX networking (1982) and developing a hybrid shell programming technique (1982) that was a precursor to the PERL programming language. He was also the manager of the AT&T ten-year future products development database. His work helped lead to commercial UNIX on computers such as Sun Micros, IBM AIX minis, and Apple computers.

In the 1980's he pioneered the development of PC Desktop Publishing on laser printers. and was nominated for three "Awards for Technical Excellence" in 1987 by PC Magazine for PC software products that he designed and developed.

In the past ten years Dr. Blaha has written over 35 books on a wide range of topics. Some recent major works are: *From Asynchronous Logic to The Standard Model to Superflight to the Stars, All the Universe!* and *SuperCivilizations: Civilizations as Superorganisms.*

www.ingramcontent.com/pod-product-compliance
Lightning Source LLC
Chambersburg PA
CBHW061417210326
41598CB00035B/6248
9780984553099